What is the time scale for the settlement and cracking of an old stone building? How do the elegant flying buttresses of a Gothic cathedral safely transfer thrust to the foundations? And what is the effect of wind on a stone spire, or bell-ringing on a church tower? These and many other questions pertinent to the upkeep of old stone structures are answered in this examination of the structural action of masonry.

With a firm scientific basis, but without the use of complex mathematics, the author provides a thorough and intuitive understanding of masonry structures. The basis of the analysis is introduced in the first two chapters, after which individual elements — including piers, pinnacles, towers, vaults and domes — are considered in more detail.

This informative text will be of particular interest to structural engineers, practising architects and others involved in the renovation and care of old stone buildings.

The Stone Skeleton

The Stone Skeleton

Structural Engineering of Masonry Architecture

Jacques Heyman

Emeritus Professor of Engineering, University of Cambridge

CAMBRIDGE
UNIVERSITY PRESS

PUBLISHED BY THE PRESS SYNDICATE OF THE UNIVERSITY OF CAMBRIDGE
The Pitt Building, Trumpington Street, Cambridge CB2 1RP, United Kingdom

CAMBRIDGE UNIVERSITY PRESS
The Edinburgh Building, Cambridge CB2 2RU, United Kingdom
40 West 20th Street, New York, NY 10011-4211, USA
10 Stamford Road, Oakleigh, Melbourne 3166, Australia

© Cambridge University Press 1995

First published 1995
First paperback edition 1997

A catalogue record for this book is available from the British Library

Library of Congress Cataloguing in Publication data

Heyman, Jacques.
The stone skeleton : structural engineering of masonry architecture / Jacques Heyman.
p. cm.
Includes bibliographical references and index.
ISBN 0 521 47270 9
1. Building, Stone. 2. Structural analysis (Engineering)
3. Plastic analysis (Engineering). I. Title.
TA676.H49 1995
693'.1–dc20 94-34084 CIP

ISBN 0 521 47270 9 hardback
ISBN 0 521 62963 2 paperback

Transferred to digital printing 1999

Contents

Preface

'The Stone Skeleton' was published as an article in the *International Journal of Solids and Structures* in 1966. That paper explored the mode of action of masonry construction, using the principles of plastic design developed originally for steel frames. The principles were applied to the analysis of the structural system of the Gothic cathedral, and the flying buttress and the quadripartite vault were treated in some detail, together with a brief discussion of domes. The paper attempted, in part, to treat the main elements in masonry construction, but there were very large gaps in the study. Other papers followed, filling in some of the holes, on spires and fan vaults, for example, on a fuller discussion of domes, and on the mechanics of arches. Some of the analytical work was collected and published in book form: *Equilibrium of shell structures*, Oxford, 1977, and *The masonry arch*, Ellis Horwood, 1982.

This present book attempts a synthesis of these studies of masonry, and presents a view of structural action which, it is hoped, will be helpful to those who wish to understand how a particular stone building might behave. Numerical calculations are made when they are necessary for the exposition, but there is virtually no mathematics in the present text; the analytical background may be found in relevant papers quoted in the bibliography. In particular, some attention is paid to the pathology of the different structural forms, and this is, of course, relevant to the repair and maintenance of a building. Since maintenance and repair, and more major works which might perhaps be classed as restoration, lead to a view of the masonry structure which lies outside the purely engineering field, it may be as well to discuss some of the considerations involved.

The archetypal problem of repair, which seems to have no easy solution, may be stated as follows. A medieval ashlar wall, perhaps the wall of a tower, has lost part of its thickness during the centuries; it is no longer a plane surface, and has indeed acquired a visual interest that was lacking in the original construction. One stone in the wall, however, originally imperfect or perhaps wrongly set by the builder, has eroded to the point where all are agreed that it must be replaced. Should the new stone be shaped to the original building line, perhaps a centimetre or so in front of the present surface of the wall? This would be disturbing visually, and in any case would lead to problems with the new stone, whose exposed edges would be subject to severe environmental attack. Should, then, the new stone be dressed back to blend in with the overall weathered surface of the original wall, so that the casual eye cannot detect the repair? Contrariwise, should a repair of this sort be distinguishable immediately to the casual eye, by the use of some different colour or other means?

Different people have different answers to these questions. The most severe view attaches paramount importance to the retention of all the material of the original structure — if, then, necessity dictates the replacement of a stone, that stone must not be 'faked' to look as if it were original. A more relaxed view would acknowledge the importance of the integrity of the original structure, but attach no particular virtue to individual stones. Rather, the quality of the wall lies in its shape and colour, and relationship to other elements of the building; there is an essential skeleton, which can be clothed with a skin which is subject to the many ravages of time, but these ravages will not affect the essence of what lies beneath.

Argument between the proponents of such severe views on the one hand and relaxed on the other can occur even over the replacement of a single stone in a large wall. The argument becomes sharper when, for example, a complete flying buttress must be taken down and rebuilt in new stone. However, from the point of view of this book, the old and the new flying buttress are identical. Unless the new work is deliberately made to assault the eye, then the old and the new will look the same, and the mode of action, whatever the looks, will actually be the same. It is precisely with the structural action of the stone skeleton that this book is concerned.

1
Introduction

It is, perhaps, trivial to remark of Greek, Roman, Byzantine, Romanesque and Gothic buildings that some of them still exist. The observation has force, however, when placed in a structural context. A masonry structure – a cathedral from the High Gothic period, for example – may be viewed in many ways: from the liturgical aspect, or the cultural, the historical, or the aesthetic, all of which may give rise to disputes of one sort or another. There remains one viewpoint which seems to engender an unequivocal statement: the large masonry building is clearly a feat of structural engineering. Moreover, the mere survival of ancient buildings implies an extreme stability of their structure.

Minor failures have, of course, occurred, and there have been major catastrophes. The fact remains that two severe earthquakes only slightly damaged Hagia Sofia, and the bombardments of the Second World War often resulted in a medieval cathedral left standing in the ruins of a modern city. At a much less severe level of disturbance, the continual shifts and settlements of foundations experienced over the centuries seem to cause the masonry structure no real distress, although, as will be seen, there may be an initial high-risk period of about a generation after completion of the building. It is the intention of this book to explain this extraordinary stability. A discussion of the actual structural behaviour of masonry is necessarily involved, and some of the history of structural analysis will be touched on, since it may help to deepen understanding.

The general principles which will be established apply to any form of masonry construction, but they will be developed by reference to the simple voussoir arch, whose behaviour is particularly easy to describe. Examples will be drawn from Gothic, because it is in Gothic that the problems of structural engineering applied to masonry are encountered in their most critical form. Only domes from the pre-Gothic period (and

1

running through Gothic into Renaissance) present any structural problem not otherwise met in Gothic itself. Indeed the collapse from Gothic to Renaissance (an aesthetic view which may perhaps be disputed) is reflected in the absence of structural interest in any Renaissance building (an engineering view which is unequivocal). It is not until the end of the nineteenth century that the gradual introduction of iron and then steel, to be followed by reinforced concrete, prestressed concrete and shell structures, has led to a renewed interest in structural analysis.

A byproduct of the structural examination of masonry is the light thrown on the activities of medieval architects. A Gothic cathedral was designed by a man who was both architect and engineer – or, of course, by a succession of such men, if the building campaigns spread over decades. The 'master of the work' had survived the long training of apprentice to journeyman to the career grade of master, and had been one of those few outstanding masters who were put to school again in the design office, before finally achieving control of a major work. This educational path contrasts strongly with that of modern Western European practice, which is based upon the Renaissance concept of the 'gentleman' architect, for whom considerations of history and aesthetics are divorced to some extent from those of engineering structure. If the building is complex, then the architect must work hand in hand with a technical adviser. The evolutionary tree of the modern structural engineer has its roots in Gothic, and earlier; that of the modern architect in Renaissance.

There was not this division in the thirteenth century (or in the sixth, when Justinian employed two outstanding Greeks, Anthemios and Isidorus, to design Hagia Sofia). The architect then knew, in the fullest technical sense, how to build, as well as how to give his building an 'architectural' design. The question seems to be obscured by the technical evidence – for example, that there are immense differences in the visual aspects of each one of the great ring of High Gothic cathedrals round Paris. Or again, the records of the late Gothic expertise at Milan (of which more will be said in Chapter 8) led Ackerman (1949) to the tempting conclusion ' ... that structure plays a secondary role in the process of creation'. This conclusion may just be true for part of the work at Milan; once it is believed to be true for any cathedral, then it is clear that the views of the 'architect' override those of the 'engineer', and Gothic is being thrust aside by Renaissance. Harvey (1958) has discussed this: 'The Gothic rules were so complicated that no one who had not served a long apprenticeship and spent years of practice could master them; whereas the rules of Vitruvius were so easy to grasp that even bishops could understand them, and princes could try their hand at design on their own.'

Vitruvian rules, however, find no place for the flying buttress or for the rib vault. These two structural elements, which may perhaps be thought to represent the essence of Gothic (see e.g. Choisy 1899), would seem to demand a long apprenticeship indeed for the mastery of their design. The medieval rules, the secrets of the lodges, ensured that the structure was effective; 'decorative' developments could then take place safely. The rules shaped the skeleton, and were themselves subject to evolutionary change; the skeleton once fixed, however, could be fleshed in a wide variety of forms.

To take a single example, the great range of vaults, from the simple quadripartite through the lierne to the fan vault, have a skeletal 'shell' very much in common, and their basic structural action is the same. This basic action stems from the nature of masonry as a material, and to understand the action it is necessary to construct a structural theory which incorporates the curious properties of masonry. Above all, it is necessary to state clearly the question that is being posed when an engineer undertakes a structural analysis: what is the problem that requires solution?

1.1 Structural criteria

In recent years structural design has come to be viewed in terms of limit states, and indeed the use of these ideas does remind the engineer that a structure must satisfy several of perhaps a large number of criteria. For example, a limit to permitted corrosion, or a restriction of crack width, may play leading roles in the design of a steel or concrete frame, respectively. These two particular criteria may also play a part in the design of masonry, although it seems reasonable to suppose them to be of secondary importance, to be reviewed finally by the designer but not likely to dictate the design.

The three main structural criteria are those of strength, stiffness and stability. The structure must be strong enough to carry whatever loads are imposed, including its own weight; it must not deflect unduly; and it must not develop large unstable displacements, whether locally or overall. If these three criteria can be satisfied, then the designer can run through a check list of secondary limit states to make sure that the structure is otherwise serviceable.

An immediate, and paradoxical, difficulty arises when the ideas of strength, stiffness and stability are applied to masonry. Ancient structures – the Roman Pantheon, for example, or a Greek temple – seem intuitively to be strong enough; they are still standing, and evidently the loading (self-weight, wind, earthquake) has not, over the centuries, caused failure to occur by fracture of the material. This matter will be discussed further, but it is a fact that mean stresses are low in a typical masonry structure; cracking and local spalling may be seen, but these seem hardly to affect the structural integrity of the whole.

Similarly, the engineer is unlikely to worry, in the first instance, about unduly large working deflexions of the vault of a Gothic cathedral. Strength and stiffness do not lie in the foreground of masonry design. Further, the engineer usually encounters instability as a local phenomenon; a slender steel column must be designed not to buckle, for example, whereas the masonry pier in a nave arcade is 'stocky'. Nevertheless, it is the third major criterion, that of stability, that is relevant for masonry, albeit in a curious form. As an illustration, the masonry arch of fig. 1.1 may be perfectly comfortable under the action of its own weight and at a certain intensity of the superimposed load P. The stresses are low and the deflexions negligible, and both will remain so as the value of P is increased. However, at a certain value of P a sudden change puts an end to this stability. As will be seen, a point is reached at which the structural forces can no longer be contained within the arch; stresses remain low but an unstable mechanism of collapse is formed (the four-bar chain of fig. 1.1(b)).

The semicircular arch of fig. 1.1 will carry a given load P provided that the arch ring has a certain minimum thickness; the design of the arch consists in the process of assigning this thickness for a given span and a given load. Once the design has been made and the arch constructed, then it will at once, and ever after, satisfy the criteria of strength (it will not crush), of stiffness (deflexions will be negligible) and of overall stability (a four-bar chain will never develop). The design consists, somewhat strangely to the mind of the modern engineer, in assigning correct proportions to the arch.

Such a design process would not have seemed strange to an ancient or to a medieval designer. It was precisely rules of proportion that were used by classical builders to design their structures, as is at once evident, for example, from Vitruvius. The brief historical notes in Chapter 8 show that these rules were never lost; they survived the Dark Ages, built in to the secret books of the Masonic lodges, and flourished in the twelfth and thirteenth centuries in the age of High Gothic. Moreover, it will be

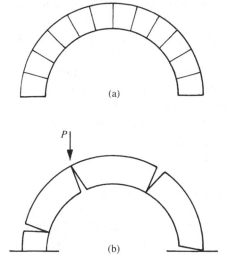

Fig. 1.1. Collapse mechanism for a masonry arch.

seen that rules of proportion give a fundamentally correct understanding of the design and behaviour of masonry.

1.2 Modern analysis

Rules of proportion, then, will be seen to lead (when correctly applied) to a masonry structure that will stand up. No statement is made about any margin of safety (if, indeed, any meaning could be attached to such a statement) nor about the loads which might cause collapse. Ancient and medieval designers did not apparently ask such questions, although they would have been well aware of structural failures – Anthemios and Isidorus only succeeded at their third attempt at a satisfactory design for the dome of Hagia Sofia.

It was Galileo who first considered the analysis of the strength of a structure, in one of his *Two new sciences* of 1638, and thereby signalled the end of medieval structural theory. He posed the problem of the assessment of the strength of a cantilever beam; what was the value of the breaking load? This sort of question may have been asked before (a tree trunk across a ditch will break under a heavy load), but not until Galileo in the context of the development of practical design rules. He wished to determine the strength of the transversely loaded beam as a function of its breadth and depth, so that a formula could be derived from which the strength of any other (rectangular cross-section) beam could be calculated. This is an example of what a modern engineer recognises as the design process (for strength of the structure).

Galileo solved the problem, essentially correctly, and found that geometrical rules of proportion no longer applied; if the dimensions of the beam were doubled, the strength was very much more than doubled. The new science of structural mechanics was eagerly pursued in the eighteenth century, and the idea of stress slowly emerged; two centuries after Galileo, the problem of the breaking load of a cantilever beam had been transformed into a problem of the determination of the value of stress in that beam. Side by side with these advances through rational mechanics, experiments on commonly used building materials had established reference values of limiting stresses. It was a natural step to attempt to relate the two values, to try to arrange that the calculated working values of stress should have an adequate margin of safety when compared with the known limiting values for the materials used.

Navier (1826) seems to have been the first to declare that the engineer was not in fact interested in the collapse state of the structure (that is, in answering Galileo's question), because all were agreed that it was a state to be avoided. Rather, Navier effectively reasserted the medieval requirement that the building should stand. But this was now to be assured not by assigning certain geometrical proportions to the structure, but by the calculation of stresses throughout its elements. It is, implied Navier, the engineer's job to calculate the actual, or working, state of the structure, and to ensure that the associated stresses do not exceed a safe fraction of their ultimate values.

Prima facie this seems a sensible procedure, but doubts arise when the analytical process is examined in detail. The designer must as a first step find the internal forces in the structure, so that corresponding values of the stresses may be calculated. The first equations written are those of statics; the internal forces must be in equilibrium with the external imposed loads. If these equations can be solved straight away then the first step is complete (and, technically, the structure is statically determinate). Generally, however, the equilibrium equations, standing alone, are insoluble; the structure is statically indeterminate (hyperstatic). There are many possible equilibrium states, that is, there are many ways in which the structure can carry its loads, and other information must be introduced into the analysis in order to determine the actual state.

Before examining this other information, however, it may be noted

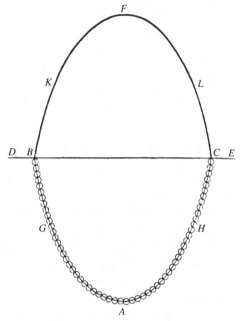

Fig. 1.2. Hooke's hanging chain.

that the arch of fig. 1.1 has infinitely many equilibrium configurations. Robert Hooke had concerned himself with 'the true Mathematical and Mechanical form of all manner of Arches for Building', and he published an anagram in 1675 (inserted in a book on helioscopes) which, rightly read and translated from the Latin, gives the statement: 'As hangs the flexible line, so but inverted will stand the rigid arch.' Hooke was unable to add the mathematics to this powerful theorem, which is illustrated in fig. 1.2. This sketch is due to Poleni (1748), whose work on the dome of St Peter's is discussed later; the shape of the chain hanging in tension loaded by its own weight is the same as that of the arch which will carry the loads in compression.

The geometry of this thrust line, that is, the actual shape of the ideal arch to carry the specified loads, will depend on the length of the equivalent chain and the distance apart of the supports. Thus a possible inverted chain is shown in fig. 1.3 lying within the boundaries of the semicircular arch; this represents one of the infinite number of ways in which the arch can carry its own weight. For masonry, as will be seen, thrusts must lie within the boundaries of the construction, and it is clear

Fig. 1.3. Line of thrust in a circular arch.

that many other catenaries could have been drawn between the extrados and the intrados of the arch of fig. 1.3.

The equations of equilibrium, used alone, do not give enough information to determine the actual position of the thrust line in fig. 1.3. The other two statements of structural analysis must be used. The first of these is a statement of material properties, so that internal deformations of the arch may be related to the internal forces. The second is a geometrical statement, and usually involves some internal or external constraint on the structure; in fig. 1.3, for example, the arch rests firmly on rigid foundations, and the internal deformations of the arch must be such that they are compatible with such imposed boundary conditions – in this case, that the displacements of the arch are zero at the abutments.

1.3 The elastic solution

Navier's design philosophy involves, then, the postulation of an (elastic) law of deformation and the assumption of certain boundary conditions that arise in the solution of the problem. If the arch of fig. 1.3 is supposed to be completely rigid (as, for all practical purposes, it is) then the position of the thrust line cannot be calculated. If, however, the arch is allowed to deform slightly (and the essence of the theory of structures is that it deals with the mechanics of slightly deformable bodies), then enough equations can be made available to solve the hyperstatic problem. If the material of the arch of fig. 1.3 is linear-elastic (or obeys any other known deformation characteristic), and if the abutments are rigid, then a unique position can be calculated for the thrust line which equilibrates the given loads (self-weight and any others specified).

It is at this point that doubts about the procedure arise. Examination of the equations shows that their solution, for a hyperstatic structure, is extraordinarily sensitive to very small variations in the boundary conditions. If one of the supposedly fixed abutments of the arch in fig. 1.3 should suffer a small displacement, this would be accompanied by a large shift in the position of the thrust line. (Small in this context implies a displacement containable within the thickness of the lines of fig. 1.3. The eye would detect no difference between drawings of the arch in the originally perfect and in the displaced states.)

Now it is certain that one or both of the abutments of the arch will in fact suffer (unpredictable) small displacements, and this then puts in question the whole of this analytical procedure. The 'actual' state can indeed be determined, but only by taking account of the material properties (which may not be well-defined for an assemblage of say stones and mortar), and by making some assumptions about compatibility of deformation – for example, the boundary conditions at the abutments of the arch. Even then, it must be recognized that the 'actual' state of the structure is ephemeral; it could in theory be determined if all the conditions affecting the solution were known exactly, but a severe gale, a slight earth tremor, a change in water table will produce a small change in the way the structure rests on its foundations, and this will produce an entirely different equilibrium state for the structure.

An analogy may help to clarify the discussion. The forces in the legs of a three-legged table are statically determinate; three equations of equilibrium may be written from which the forces in the legs may be evaluated to support a given weight placed in a given location on the table. The addition of a fourth leg makes the problem very much more difficult, since the table is now hyperstatic; the same three equations may be written, but four forces have to be evaluated. The solution to the problem requires a knowledge of the flexibility of the table top, details of its connexion to the legs, the compressibility of the legs themselves, and so on. A computer program could perhaps be used, and the four required values of the forces in the legs can be determined. However, hidden in this program, and perhaps not noticed by the user of the program, there will be implicit boundary conditions; it will be assumed, for example, that the table is standing on a rigid level floor.

Now a real stiff table on a rigid floor will rock; if one leg is clear of the ground by only one millimetre, the force in that leg will be zero, and the forces in the other three legs are then uniquely determined. A moment later the table may be jolted and will shift to another part of the floor; the table is the same, and it is carrying the same load with comfort, but a different leg is now off the floor, with corresponding unique forces in the remaining three. Both of these states of the table are possible equilibrium states, as indeed is the computer-generated state, but none represent the 'actual' state.

1.4 Plastic theory

These conclusions are not academic abstractions. A series of tests carried out by the Steel Structures Research Committee in the early 1930s showed that the stresses (actually strains) measured in practice, in office blocks and hotels, for example, bore almost no relation to those confidently calculated by the designers. Moreover the Committee concluded that there was a problem incapable of solution – practical imperfections of construction and behaviour were inevitable, and would always lead to an unpredictable working state of a structure (in this case, a steel frame). What had gone wrong was the attempt to base the design of a structure on a knowledge of its 'actual' state. If progress were to be made, then this philosophy of design would have to be abandoned.

Common sense would seem to indicate that a severe gale, easily survived but leading to a completely different state of the structure, cannot really have weakened the structure – the four-legged table, accidentally knocked by the waitress, continues to serve its function. Common sense is in this case supported by both theory and experiment. If two seemingly identical structures, but actually with different small imperfections so that they are in very different states of initial stress, are loaded slowly to collapse, then the collapse loads (that is, the strength of the structures) will be found to be the same. It was this observation which led to the development of the so-called plastic theory of structures, applicable to any case where collapse is a ductile quasi-stable (plastic) process. The theory applies therefore to steel and to reinforced-concrete frames and, as will be seen, to masonry, or to any building type using a structurally common material (timber, wrought iron, aluminium alloy), but not to materials like cast iron or glass, which are brittle.

Thus the plastic designer abandons the quest for the actual state of a structure and, instead, examines the way in which that structure might collapse. It is not envisaged that the structure will actually collapse, however. Rather a calculation is based upon loads increased by a hypothetical factor; the master theorem of plastic analysis, the key tool of the designer, then states that the real structure acted upon by the smaller working loads will never collapse.

In the course of the calculation of hypothetical collapse the plastic designer generates an equilibrium state for the structure under its real working loads. Elastic designers believe that they have generated the 'actual' equilibrium state, whereas plastic designers know only that they have generated one particular state out of the infinitely many that are possible. Thereafter both types of designers will proceed, effectively in the same way, to provide their final designs of their structures with an adequate margin of safety.

What both designers are doing, one consciously and one unconsciously, is to apply the master safe theorem of plasticity. If any equilibrium state can be found, that is, one for which a set of internal forces is in equilibrium with the external loads, and, further, for which every internal portion of the structure satisfies a strength criterion, then the structure is safe. In conventional terms, the strength criterion might be that the stresses at every cross-section of the structure are less by some margin than the yield stress of the material; in terms of masonry, the criterion is effectively that the forces should lie within the boundaries of the material. The power of this safe theorem is that, in either case, the equilibrium state examined by the designer need not be (indeed, from what has been said, cannot be) the actual state; viewed anthropomorphically, the safe theorem declares that, if the designer can find a way in which the structure behaves satisfactorily, then the structure itself certainly can. A simple exposition of this theorem is given in the next chapter with reference to the voussoir arch.

This, then, is the philosophical framework within which the structural action of masonry will be pictured. A detailed examination of the properties of the material enable some powerful simplifications to be made, leading to broad statements concerning the behaviour of individual elements of a building.

2

Structural Theory of Masonry

Masonry is an assemblage of stones – or bricks, or indeed sun-dried mud (adobe) – classified for convenience with certain distinct labels, as Byzantine, Romanesque, Gothic, but recognized by engineers as having a common structural action. This action arises directly from the properties of the material.

It is prudent and convenient to regard a masonry building as a collection of dry stones (or bricks etc.), some squared and well fitted, some left unworked, and placed one on another to form a stable structure. Mortar may have been used to fill interstices, but this mortar will have been weak initially, and may have decayed with time – it cannot be assumed to add strength to the construction. Stability of the whole is assured, in fact, by the compaction under gravity of the various elements; a general state of compressive stress exists, but only feeble tensions can be resisted.

All this must have been well understood by medieval cathedral builders, although they would not have had numerical concepts of stress or of the strength of their material. A modern engineer would perhaps make calculations to relate these quantities. In this connexion reference to nineteenth-century practice in the design of great masonry arches is illuminating. An indirect parameter was used to express the strength of stone – the height to which a prismatic column might (theoretically) be built before crushing at its base due to its own weight. This height may be predicted easily; a medium sandstone might have a unit weight of 20 kN/m^3 (a mass of 2000 kg/m^3) and a crushing stress of 40 MN/m^2 (400 kg/cm^2). Dividing one figure by the other gives the height of the self-crushing column as 2 kilometres.

Yvon Villarceau used this parameter in his extensive digest of 1854 on the 'state-of-art' of masonry bridge construction, and he advocated a factor of 1/10 on the height of the column. That is, he suggested that

12

the nominal stresses should be limited to 1/10 of the crushing stress of the material, and, for the example of the sandstone above, the height of the column would then be 200 m.

In applying these ideas to a tall Gothic cathedral (say Beauvais, whose height beneath the stone vault is about 48 m), it should be noted that the stresses are almost entirely due to the vertical dead weight of the material. The nave and choir piers must in addition support the weight of the high vault and the timber roof above, and they will be subjected also to loads from wind, and perhaps earthquake. Specially widened columns at the crossing, or at the west end, may support towers, but it is difficult to imagine that even a crossing pier will be subjected to a high level of stress compared with the potential crushing stress of the material. Indeed Benouville, in his 1891 analytical study of the structure of Beauvais, astonished himself by not being able to find any stress greater than 1.3 N/mm^2. This compares with the previously mentioned crushing stress of say 40 N/mm^2; there is a large factor of safety of 30 against crushing of the material. In summary, and in very round figures, the most highly stressed elements of a masonry structure (crossing piers in a cathedral and the like) will have average compressive stresses not more than one-tenth of the crushing stress of the material. (The validity or otherwise of using 'average' stresses is discussed later.) The main portions of the load-bearing structure of a church (flying buttresses or the webs of masonry vaults, for example) will be working at one-hundredth of the crushing stress, and infill panels or walls which carry little more than their own weight may be subject to a 'background' stress as low as one-thousandth of the potential of the material. The enormous factors of safety against crushing implicit in these figures make their rough derivation unimportant.

A low background of compressive stress is essential for the stability of the masonry structure. The small pieces of stone are compacted by gravity into a certain overall shape designed by the architect, but that shape can only be maintained if the stones do not slip one on another. The elements may interpenetrate to some extent, or may be cut with some joggles and keys; in the absence of such designed roughness, the main instrument of stability is the low compressive stress which will allow friction forces to develop, locking the stones against slip. (It is evident that the stones themselves must have a certain minimum size; a dry stone wall can stand, but an attempt to build the wall from sand would be unsuccessful – the sand would slump away.)

Thus the behaviour of the masonry structure can be examined in the light of three simplifying assumptions, each one of which is not strictly true and must be hedged with qualifications, and which must in any case be tested in the light of contradictory experience with a particular building. The three assumptions are that

(i) masonry has no tensile strength,

(ii) stresses are so low that masonry has effectively an unlimited compressive strength, and

(iii) sliding failure does not occur.

The first assumption is clearly conservative, but not unduly so. Individual blocks of stone may be strong in tension (and corbelled construction relies on this), but mortar between stones is indeed weak. An attempt to impose tensile forces would pull the work apart.

The assumption of unlimited compressive strength of the material will be approximately correct if *average* stresses are in question. Stress concentrations can arise, however, which will lead to distress evidenced by splitting or surface spalling. Such distress is local, and will not normally lead to overall failure of the building, but the assumption must certainly be questioned in relation to the behaviour of apparently solid walls, which may actually consist of two skins containing a rubble fill. The matter is discussed further in Chapter 5.

Finally, there is sometimes evidence to be seen of slippage of individual stones, but generally the masonry structure retains its shape remarkably well; evidently a very small compressive prestress is all that is necessary to avoid the dangers of slip and general loss of cohesion. Occasionally, particular precautions must be taken to avoid slip, and a discussion is given in Chapter 5 of the action of pinnacles on buttresses, but in general it is rare for a masonry structure to be distressed in this way.

The three assumptions are in fact those required to give an account of the action of masonry which lies within the general field of plastic theory. The basic theorems of plasticity are then available, and it is possible to make powerful and effective general statements about the behaviour of the large masonry structure. For simplicity, the principles will be presented with reference to a single-span voussoir arch, but the conclusions apply to any, and much more complex, form of masonry construction, for example a complete Gothic cathedral.

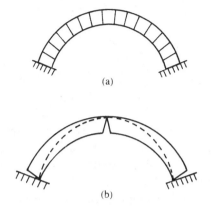

Fig. 2.1. Stable state of a cracked voussoir arch.

2.1 The voussoir arch

The arch of fig. 2.1(a) is supposed to be made from identical wedge-shaped stone voussoirs, assembled without mortar on temporary formwork (centering), and fitting exactly between its abutments. The centering supporting the masonry is then removed, and the arch starts to thrust horizontally. The abutments, inevitably, give way slightly by unknown amounts. What meaningful statements can be made, under these conditions, about the structural state of the arch?

Any structural theory must take account of the three structural statements noted in Chapter 1. First, the equations of equilibrium must certainly be satisfied; the internal forces in the arch must equilibrate the external loads (including self-weight). Second, any statement about behaviour of the material will accord with the three simplifying assumptions, that is: the material is infinitely strong in compression, cannot accept tension, and slip does not occur. The voussoirs of the arch will in fact be taken to be rigid blocks, not deforming in themselves, and there is thus no possibility of obtaining any 'elastic' solution to the state of the arch. Finally, it will be accepted that no statement can be made about the boundary conditions; the movements of the abutments will be considered to be unknown.

If in fact the abutments in fig. 2.1 do give way by a small unknown amount there will be a small geometrical mismatch; if the arch does not fall, it must somehow accommodate itself to a slightly increased span. This accommodation must be made by cracking of the arch, as shown (greatly exaggerated) in fig. 2.1(b). The voussoirs cannot slip, and they cannot deform within themselves, but they can turn about points of contact with each other either on the extrados or on the intrados; the resulting cracks, which may be hairline, or even closed in practice by the real elasticity of the stone, may be idealized as hinges. Whether the cracks are idealized or not, they will always occur in a real stone structure. For the arch, the abutments will spread; for a complex masonry structure, foundations will settle and mortar will dry and shrink; in all cases a slightly different geometry will be imposed, causing the masonry to shift. The simple voussoir arch has a simple pattern of cracks; the complex structure will exhibit a complex pattern, with cracks developing not only between but perhaps also through the stones.

Such cracking is inevitable and indeed may be thought of as natural; it is by no means a sign of incipient collapse. It indicates merely that there has been some unpredictable and irresistible shift in the external environment to which the structure has responded. The cracked three-pin arch of fig. 2.1(b) is in fact a well-known and perfectly satisfactory structural form. Moreover, the three-pin arch is a statically determinate structure, and the internal forces can be found purely from the equations of equilibrium. This is perhaps obvious from the position of the thrust line sketched in fig. 2.1(b); the thrust line is clearly constrained to touch the extrados of the arch at the crown, since this is the only point (at the hinge) capable of transmitting forces. Similarly, the thrust line must pass through the intrados at the abutment hinges.

Thus, in this particular instance, the problem referred to in Chapter 1 and implied by fig. 1.3 has been solved. Out of the infinite number of locations of the thrust line, one has been selected uniquely by the small unknown movement apart of the abutments. The corresponding value of the horizontal component of the abutment thrust can be calculated, and it may be noted that this value of the thrust does not depend on the magnitude of the small unknown spread of the arch. As a matter of interest, the actual value of the thrust in this configuration has its minimum possible value if the stability of the arch is to be assured; as the abutments start to spread, the thrust falls at once to this least value, and is then fixed.

A full semicircular arch has been drawn under similar conditions in fig. 2.2(a). The geometrical properties of a catenary and a circle are such that the intrados hinges now form away from the abutments, but the arch is once again statically determinate, and the minimum value of the abutment thrust may be calculated. Had the abutments moved slightly closer to each other rather than slightly apart (as perhaps they might

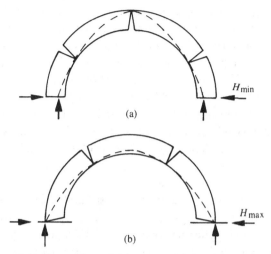

Fig. 2.2. The semicircular arch under its own weight. (a) Minimum abutment thrust. (b) Maximum abutment thrust.

should the sketch represent an internal arch of a multi-span bridge) then once again hinges will form (fig. 2.2(b)), this time under the action of a maximum value of the abutment thrust, and once again the value of that thrust is calculable immediately.

The broken lines in Figs. 2.2(a) and (b) represent the limits between which all possible positions of the line of thrust must lie. It will have been noted that when the arch is acting upon the environment (thrusting at the river banks which are compelled to give way) then the thrust falls to its minimum value; by contrast, when the environment acts upon the arch (adjacent spans thrusting at the span in question), then the arch resists to its greatest possible extent, and develops the maximum thrust of which it is capable. The behaviour shown by this simple example has its counterpart in that of the general masonry structure; as will be seen, it makes a well-designed flying buttress a very effective structural element.

Two further remarks may be made about fig. 2.2. First, the hinging crack at the crown of the arch in fig. 2.2(a) will be visible from the underside of the arch, and it is indeed a common sight when passing under a masonry bridge to see a crack running longitudinally along the barrel. This matter is raised again in Chapter 4 in the discussion of the pathology of masonry cross-vaults. Second, the thrust lines in fig. 2.2(a) and (b) are shown as passing exactly through the hinge points, as theoretically they must. However, a finite load acting through a line contact implies infinite stresses at that contact, and the first assumption relating to the material behaviour (no limit to compressive stresses) must be questioned. In practice the high stresses will cause material failure at the contact; there may be some crushing of the stone (so that the contact area becomes finite and the stress drops to an acceptable level) or spalling and splitting may occur – such spalling may sometimes be seen under the barrel of a bridge in the regions indicated by the intrados hinges of fig. 2.2(a). Again, the question of such spalling is discussed further below.

Putting these practical considerations to one side, and reverting to the idealized material (infinite compressive strength, zero tensile strength, no slip), it is possible to 'translate' basic ideas stemming from plasticity theory into terms applicable to masonry. First, the way in which the simple arch might collapse may be considered. Figure 1.1 is repeated in part of fig. 2.3; the semicircular arch is acted upon by its own weight and an additional point load P. Thus Hooke's hanging chain will be deformed from the catenary to the shape shown in fig. 2.3(b). As P is increased, the inverted chain (that is, the line of thrust) will fit less and less comfortably within the arch, and at a certain value of P the thrust line can only just be contained. This limiting stage is shown in fig. 2.3(c), and it will be seen that the thrust line reaches the surface (either the extrados or the intrados) at four locations. At each of these locations a hinge will form, and four hinges transform the stable arch structure into a mechanism of collapse, the four-bar chain of fig. 2.3(d). It will be appreciated that, if the black dots in fig. 2.3(d) represent frictionless hinges, then the configuration shown would correspond to a real mechanism.

By contrast the three-hinge arch, referred to above as a satisfactory structural form, is stable precisely because three hinges are insufficient for the formation of a mechanism. If the arch were initially in the state of fig. 2.2(a), that is, in the 'three-pin' state following slight spread of the abutments, then it may well be imagined that the application of a point load P could close some of these pre-existing cracks, and that some new cracks might open. However, none of this will lead to loss of structural integrity until the load P reaches its critical value – failure of the arch will occur only when sufficient hinges have formed (four in this case) to turn the structure into a mechanism.

In fact, collapse only occurs finally at the point when it is no longer possible to fit the thrust line within the boundaries of the arch. This is

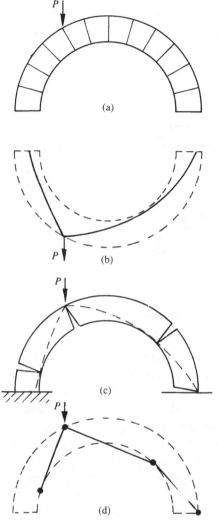

Fig. 2.3. Collapse of a circular arch under a point load.

a geometrical constraint, not related to any initial state of the arch –
whatever its initial crack pattern, the final mode of collapse is represented
by fig. 2.3(c). This is an example of the 'common-sense' view that
small initial imperfections, while leading to different initial states of the
structure, do not affect its ultimate strength.

It may be noted in passing that there are some forms of arch for which

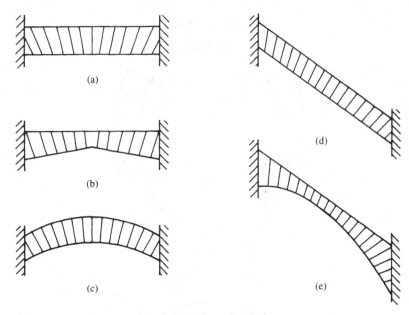

Fig. 2.4. Various flat arches.

a mechanism of collapse cannot be constructed. There is no arrangement of hinges in the extrados and intrados of the flat arch of fig. 2.4(a) (the *plate-bande*) which will give rise to a mechanism, and the same is true of each of the arch forms sketched in fig. 2.4. The conclusion is that, within the present assumptions, these types of arch are infinitely strong. They will fail only when the loads they carry increase to such an extent that overall crushing occurs (local crushing may be tolerated), or alternatively by one of the stones sliding out of the construction. The arch form of fig. 2.4(e) will be discussed in more detail when the function and mode of action of the flying buttress is examined in Chapter 5.

2.2 The geometrical factor of safety

The thrust line of fig. 1.3, redrawn in fig. 2.5(a), represents, as has been seen, only one of the infinite number of possible states of equilibrium of the arch. The physical arch in fig. 2.5(a) has, in fact, sufficient depth to accommodate a range of thrust lines arising from the self-weight of the material of the arch. If this self-weight is distributed uniformly round the arch, then the shape of the thrust line is that of the mathematical

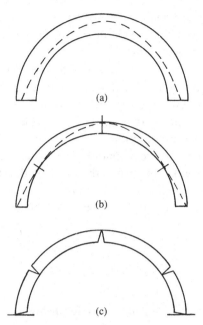

Fig. 2.5. The semicircular arch (a) stable and (b), (c) of minimum thickness.

catenary (as sketched), and it will be appreciated that there is a minimum thickness of semicircular arch which will just contain a catenary. Such an arch (of thickness just over 10 per cent of the radius) is sketched in fig. 2.5(b). A thinner arch cannot be constructed without the line of thrust passing outside the masonry, which would imply tension in the material in contradiction to the no-tension assumption. The arch of fig. 2.5(b) is on the point of collapse by the formation (theoretically) of the five-bar chain of fig. 2.5(c); any slight asymmetry (of geometry or loading) will suppress one of the abutment hinges and reduce the mechanism to the regular four-bar chain.

The arch of fig. 2.5(b) is containable within the arch of fig. 2.5(a), and this leads to the idea of a geometrical factor of safety, defined by the ratio of the thicknesses of the two arches. If this factor were 3, for example, then the thrust line for the arch of fig. 2.5(a) could be contained within its middle third (the 'middle-third rule' was in fact originally an elastic concept concerned with the avoidance of tension). If the factor were 2, then the thrust line could be contained within an arch of half the depth, and so on. (As a matter of interest, safe practical values for the geometrical factor of safety in arch construction seem to be about 2;

such a factor will guard against asymmetries engendered by construction or subsequent settlement, and will allow also for accidental or otherwise unforeseen superimposed loading.)

The idea of a geometrical factor of safety in masonry construction therefore replaces the conventional idea of a strength factor in the design of usual (say steel or reinforced-concrete) building structures. In masonry, for which the assumption has been made that the material is infinitely strong, there is in fact no question of an overall criterion of strength (it must be remembered that local conditions – spalling and the like – may have to be investigated). Rather it is the shape of the structure that must be examined; certain minimum dimensions must be given to the elements and to the overall structure so that thrusts may be accommodated within the material.

2.3 The master 'safe' theorem

The discussion of the geometrical factor of safety has already accepted implicitly a certain view of the behaviour of masonry. With reference to fig. 2.5, it is that if the arch of fig. 2.5(b) can be fitted within the boundaries of the actual arch of fig. 2.5(a), then the actual arch is indeed safe.

All this can be put more formally. The arch must be in equilibrium, and the mathematics of equilibrium can be represented by Hooke's inverted chain, that is, by the thrust line. The material requirement (no tensile forces) requires the thrust line to lie within the masonry. The safe theorem states that if any one such position can be found for the line of thrust, then this is an absolute proof that the structure is stable, and indeed that collapse can never occur under the given loading.

As noted in Chapter 1, the power of this theorem lies in the fact that only one position need by found – no reference is made to the 'actual' state. It was seen that the actual state is in any case ephemeral; small changes in the structure's connexions with the environment can cause large changes in the position of the thrust line. However, to continue the anthropomorphic view of Chapter 1, once it has been demonstrated that the thrust line *can* lie within the masonry then, no matter how wildly it thrashes about in response to shifts and settlements, it can never escape.

Some simple lemmas will make clear the power of the safe theorem.

2.4 Settlement and cracking

The abutment spread of a masonry arch was imagined to lead to a tidy set of hinges; in practice, it might be difficult to identify resulting cracks in the real masonry arch with such a simple theoretical pattern. Settlements in a highly complex structure, say of a nave pier relative to its neighbours in a cathedral, will at first glance seem almost random. Interpretation is in fact usually possible, but that is not the present point. The question is whether or not a visually alarming set of cracks, say in the masonry just above the nave pier which has settled differentially by perhaps as much as 100 mm, is in reality a sign of danger.

If an outline drawing of the nave arcade were made to a scale of 1/100, it would be accommodated on a relatively modest size of paper. Suppose two such drawings were made, the first of the arcade in its original perfect state, and the second showing the settlement of 100 mm of the nave pier. It is clear that the two drawings could then be superimposed almost exactly – the defect of 100 mm in the real structure would be represented by 1 mm on the drawing board, barely more than the thickness of a pencil line.

Now the geometry of the original arcade, before any settlement occurred, is known to have been satisfactory; the arcade stood, which is experimental evidence that a set of forces could be found lying within the boundaries of the masonry. The geometry of the distorted arcade is virtually unchanged – the force paths sketched on the original drawing will lie within the masonry of the second drawing. The conclusion from the safe theorem is that, despite perhaps visually alarming cracking, the deformed arcade has virtually the same margin of safety as the arcade in its perfect state.

Masonry is supposed to crack, and any cracks visible in a structure indicate merely that the building has at some time been subjected to imposed movements from the external environment. If there is no evidence of recent movement (but see the next section for a discussion of the soil-mechanics time scale) then the proper course of remedial action is to point cracks with mortar (to prevent ingress of water which may lead to more rapid internal deterioration), but otherwise to leave well alone. Indeed, in a figurative sense, if waterproof paper is available, then a thorough repair of settlement damage may be effected by papering over the resulting cracks.

2.5 Time scale for settlement

The five-minute theorem for masonry may be stated for, say, a flying buttress. If the timber centering of the flying buttress is removed on completion of the stonework, and the work stands for 5 minutes, then it will stand for 500 years. The 5 minutes is the time needed to confirm, experimentally, that the shape of the flying buttress is correct; evidently the boundaries of the masonry contain the thrust line, and the safe theorem is satisfied. The upper limit of 500 years depends on the decay of the material.

This flamboyant statement assumes that the loads on the flying buttress are static, arising from self-weight and dead thrust from the nave vaulting. The statement does not necessarily apply to the upper flying buttresses in some large Gothic cathedrals, one of whose functions (which will be mentioned again later) is to resist wind forces acting on the great timber roof. The statement also concentrates only on the masonry, and ignores any interaction there may be with the foundations. Settlements of foundations are measured on time scales of years, even a decade or so, and not minutes.

Foundation settlements can lead to geometry changes rather larger than those envisaged so far. Examinations of crossings of cathedrals, where there are (or have been) large towers, almost always give evidence of more or less gross distortions. Initially horizontal courses of masonry in the nave, choir and transepts abutting a crossing may indicate settlements of as much as say 300 mm (or 3 mm on the drawing at 1/100 scale, now rather more than the thickness of a pencil line). This is both commonplace and straightforward; the four piers supporting the tower, themselves highly loaded (and stressed perhaps as high as one-tenth of the strength of the material), will require high bearing stresses from the soil for their support. Typically, even if the whole plan area at the crossing is assumed to be involved in carrying the tower, the mean stresses are so high that a modern engineer redesigning the cathedral would be forced to use piles to limit foundation movement. It should be remembered that the interest of the modern engineer is to prevent such movement occurring at all, whereas the original builders were prepared to allow settlement and consequent cracking. Indeed, the high stresses under the medieval structure will ensure that settlement resulting from consolidation of the soil will inevitably occur. If the settlement can be tolerated, however, the resultant stresses are not so high as to finally distress a stiffish clay.

It is to be expected then that a tower will settle in relation to the surrounding portions of the structure. Moreover the 'soil-mechanics' time scale for consolidation of a mass of soil in an area of a cathedral crossing is a generation; effective equilibrium will be reached within this period. The 'five-minute rule' for masonry considered in isolation should be replaced by a 'generation rule' for the structure as a whole, and there are in fact a large number of recorded collapses of towers within a period of 20 years from their completion, for example at Winchester, Gloucester, Worcester and Beauvais (in 1573; there was an original collapse of the cathedral in 1284, also within 20 years of the initial build).

The precise mechanisms of collapse are complex, and some discussion is given in Chapter 6 when the structure of towers is considered. The cause in general is uneven settlements, not necessarily upsetting the overall geometry of the tower itself, but causing high strains in portions of the tower or in abutting masonry. In some cases this masonry was reinforced to restore stability, sometimes with 'hidden' internal raking buttresses, as in the late-fourteenth-century tower at Worcester, occasionally spectacularly, as with the earlier strainer arches at Wells.

If a tower (or any other part of the building) does survive the initial risk period of a generation, either untouched or with reinforcement, and provided there are no changes in the general condition of the soil, such as would be caused by alterations in the level of the water table, then the tower may well be deemed to be structurally safe. However, soil conditions *can* change, and the masonry structure is then once more at risk. The collapse of the central tower at Ely, for example, occurred two centuries after its completion; the cause is now unlikely to be discovered, but it is possible that it could be associated with one of the periodical attempts to drain the Fens. Alternatively, it may be noted that the building of the Lady Chapel had started a year earlier, with necessary disturbance of the ground nearby. Similarly, the collapse at Chichester in 1861 may have been engendered by the development of the land to the south of the cathedral. However, the tower had been stable for six centuries, and the 500-year rule may have been operated – certainly Robert Willis was of the view that decay of the masonry of the piers caused the collapse.

2.6 Models

The view of the masonry structure which emerges from this discussion is one based firmly on geometry. That is, the stability of the structure will be assured primarily by its shape, and not at all (or only very marginally) by the strength of the component material. Modern stress rules play no part in a structural view of masonry. Medieval architects may have known, in a non-mathematical way, about forces, but it seems certain that they knew nothing of stress; as will emerge in the historical Chapter 8 of this book, they were concerned with the proper proportions of a structure – the height of a pier should be a certain multiple of its width, and so on. Rules like these are essentially numerical, and lead to a geometry of structure which is independent of scale.

Once the unit of measure had been established on a building site (see Chapter 8), then all individual dimensions for all parts of the building followed by simple rules of proportion. A satisfactory design could be built to any size. In particular, a model could be used not only to solve problems of construction (for example, problems of stereotomy), but also to simulate the full-scale structure. The use of models is well attested; the late fourteenth-century brick and plaster model of S. Petronio, Bologna, was over 18 m long. Such a model can be used with confidence to check the stability of the whole or of any part of the structure, since questions of stability, depending as they do on individual proportions, that is, on geometrical shape, can be checked at any scale. Or again, Brunelleschi's use of turnips, cut to shape and quickly assembled, as described by Vasari (see Jackson(1921) vol. 1 p. 56), is not unbelievable; the turnips may have been used to solve three-dimensional problems in the cutting of individual stones, but, equally, the extremely low ambient stresses of an assemblage would have allowed the structure to stand until the '500-year rule' operated – that is, until the turnips rotted.

This, then, is the medieval view of masonry, essentially correct provided that stresses are low. By contrast, questions of strength cannot be scaled in this way. Galileo was clear about this, and his account of the square-cube law was the first scientific demonstration that a scale model will always be misleadingly strong. But strength is not the prime criterion for masonry design. A cathedral built to a height of 2 kilometres will crush under its own weight; it is the fact that stresses are low in a cathedral of human scale that makes geometrical rules effective.

3

Domes

A dome is a rounded vault forming a roof over a large interior space (e.g. Hagia Sofia, c. AD 532 with a span of about 31 m; St Peter's Rome, c.1560–90, 42.5 m). The word in Italy (*duomo*) and Germany (*Dom*) has come to stand for the whole cathedral, and indeed the etymology is from the Latin *domus*, house (of God). The French use both *dôme* and *coupole* (cf. Italian *cupola*) for the vault; the English cupola is sometimes reserved for a very small domed roof, as for example on the lantern mounted on the eye of a dome proper, cf. the cross-section of St Paul's Cathedral shown in fig. 8.9.

The 'rounded vault' of the dome can take many forms. Perhaps the simplest of these is a shell of revolution, in which every horizontal section is circular; an egg in an egg-cup is a shell of this kind. The inner dome of St Paul's is roughly spherical, and has an open eye, while the main dome carrying the lantern is conical, but both are shells of revolution, as is the surface of the third lead-covered timber outer dome; all have circular horizontal sections.

The dome of St Paul's stands on a circular cylindrical drum (or at least appears to do so; as will be seen, Wren used some visual disguise to hide the structural reality). The use of the drum to elevate the dome, for example above the crossing of a cathedral, is a relatively late development (after AD 1000). Whether or not the dome is elevated, the circular horizontal section has to be made to fit the supporting polygonal structure (usually square or octagonal, but for example hexagonal at Siena). The change from circular to polygonal is achieved with pendentives, a Byzantine invention. The need for such pendentives is obviated if the dome itself is polygonal. Brunelleschi's dome at Florence (c. 1420), for example, is octangular at every horizontal section, and is supported from an octangular drum which in turn covers an octagonal space within the cathedral.

The structural action of these different kinds of dome may be analysed using the ideas of the previous chapters, that is, using the ideas of statics (Hooke's hanging chain) applied to masonry as a no-tension material. It will be seen that, whatever their shape, all masonry domes thrust out against their supports, and must be buttressed by those supports. Thus the dome of the Roman Pantheon, *c.* AD 120 and over 43 m diameter, which is effectively a thick concrete hemisphere with an eye open to the sky, is buttressed by the massive 6 m circular wall of the building.

3.1 The dome as a membrane

In modern engineering theory a shell is a structure which can be idealized mathematically as a curved surface (just as an arch may be idealized by its curved centre line). The thickness of the shell, which need not be constant from point to point, is thought of as being small compared with the leading dimensions of the structure. When loads act on such a curved surface, whether imposed externally or arising from the weight of the shell itself, they must be resisted by forces within the surface. In membrane theory it is assumed that the surface has no stiffness against bending, so that the forces in the shell are purely tensile or compressive (and, of course, tensile forces are inadmissible for masonry, which leads to crucial departures from conventional theory).

Thus, in two dimensions, Hooke's hanging chain has no bending stiffness, and will take up a unique shape when subjected to a set of given loads. The chain is thought of as being vanishingly thin; when inverted it represents the thrust line lying within the real thickness of the masonry arch. By contrast with the flexible chain, the behaviour of a hanging flexible membrane, made say of cloth, is markedly different. Such a membrane, in the form perhaps of part of a sphere, or of any other non-developable surface, can carry a wide range of different loadings without altering its basic shape (provided that none of the forces in the membrane tries to become compressive, which would lead to distortion of the surface by wrinkling). Similarly, a thin domical shell, made of relatively rigid material, can carry a wide range of loadings purely by the action of forces within the surface of the shell, that is without bending. Thus the shell structure, unlike the arch, does not have to be built in practice with some reasonable thickness in order to accommodate safely a range of loads, although a certain minimum thickness (which is usually very small indeed) is necessary to prevent local compressive buckling.

Table 3.1.

	Span L (m)	Thickness t (mm)	L/t
St Peter's Rome	42.5	3000	14
Hen's egg	0.04	0.4	100
Smithfield Market	68	75	900

The familiar example of the hen's egg, made of thin, stiff, fragile material, illustrates some of these points. The shell usually survives intact until deliberately broken; a standard man is incapable of breaking a standard egg between thumb and forefinger of one hand, pressure being applied along the axis of revolution. It requires the application of a high local pressure, provided perhaps by the beak of a chick or by an eggspoon, to break the shell. The pressure between thumb and forefinger is transmitted smoothly from one element of the shell to the next by the action of membrane forces.

Although it is not of immediate importance to the present discussion, it may be noted that an intact egg shell is, in a technical sense, complete; it is bounded only by the two surfaces of the shell. If the egg is cut, however, two incomplete shells are generated; incomplete shells have edges. The behaviour of incomplete shells can be different from that of complete shells; certainly the decapitated egg is more flexible than the uncut egg. A hen-produced egg is an imperfect shell of revolution, and usually rocks in a machine-produced egg-cup. With the top removed, however, the incomplete shell can be given small bending deformations so that it fits well into the cup.

The shell of a hen's egg lies, according to the numerical measure shown in table 3.1, centrally between the masonry domes of the Renaissance and the shell roofs of modern concrete technology. The term thin shell is normally applied to shells for which $R/t > 20$ (it is technically better to use the minimum radius of curvature R of the shell rather than the span L in computing the thickness ratio), and practical modern thin shells, of which Smithfield Market is given as an example in table 3.1, may have this ratio as high as 1000 or more. The hen's egg has the value of the ratio an order of magnitude lower at 100; at a further order of magnitude lower still lie the domes of St Peter's and of the Roman Pantheon.

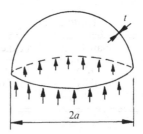

Fig. 3.1. Hemispherical shell under its own weight.

3.2 Stresses in shells

It was noted in Chapter 2 that stresses in masonry construction are in general low, and this statement applies also to the stresses in domes. To illustrate this in broad terms, fig. 3.1 shows a uniform thin-walled hemisphere subjected only to its own weight, and supported round its base (a horizontal diametral plane) by forces which produce a uniform compressive stress σ. (The way in which the forces are distributed in the shell itself is discussed in the next section.) If the shell has radius a and thickness t, then the volume of material is $2\pi a^2 t$, and it will be supposed that this material has unit weight ρ. The supporting stress σ acts on a diametral ring of area $2\pi at$, so that, equating forces,

$$\sigma(2\pi at) = \rho(2\pi a^2 t),$$
$$\text{or} \quad \sigma = \rho a. \tag{3.1}$$

Thus it will be seen that the compressive stress necessary to support the dome has a magnitude independent of the thickness; doubling that thickness will double the weight of the dome, and at the same time double the area involved in the support.

For concrete or masonry the unit weight might be 20 kN/m³, and if the dome spans the huge distance of $2a = 100$ m, then equation (3.1) gives the value of $\sigma = \rho a$ as 1 N/mm². It was noted that the crushing strength of a medium sandstone (or indeed of concrete) might be about 40 N/mm², and once again the stress level just computed is very far from the limiting stress of the material. Moreover although equation (3.1) resulted from consideration of an idealized hemisphere, the resulting stress is typical of those in more general shapes of shell, although naturally the analytical expressions may be far more complex.

Further, it may be expected that externally applied loads will be less than the self-weight loads of the shells themselves, if these are of reasonable thickness. For example, a concrete shell 100 mm thick will have a weight per unit area of 2 kN/m^2; a typical maximum snow load might be 1.5 kN/m^2, with wind loads probably smaller. Thus stresses resulting from snow or wind might be expected to be of the same order as those resulting from the self-weight of a modern thin shell roof; the effect of the weight of snow on the dome of St Peter's is negligible.

Equally, masonry domes are in no danger of buckling locally. Buckling analysis is not easy, but critical conditions arise at a typical stress σ_{cr} given by

$$\sigma_{cr} = kE\frac{t}{R} \tag{3.2}$$

where E is Young's modulus. The value of the constant k varies from author to author, but a reasonable value is about 0.25. Thus for a concrete shell for which E is about 20 kN/mm^2, and for which t/R is as small as 1/1000, the critical stress is determined as 5 N/mm^2.

3.3 The uniform hemispherical dome

Equation (3.1) gave the value of the stress necessary to support a dome in the form of a hemisphere; membrane theory will now be used to determine how the stresses are transmitted through the shell to the base. It is convenient in membrane shell analysis to work not in terms of stresses σ, but in terms of stress resultants N, where $N = \sigma t$. Similarly, self-weight loads are expressed in terms of a load w per unit area, where $w = \rho t$, so that equation (3.1) may be rewritten

$$N = wa. \tag{3.3}$$

Figure 3.2 shows the meridian of the sphere representing the dome; it has radius a. Points round the meridian are identified by the angle ϕ, which is the complement of the latitude used in terrestrial geography. A point on the meridian is distant r from the vertical axis of the sphere, and a simple geometrical relationship may be written:

$$r = a\sin\phi. \tag{3.4}$$

In order to examine the way in which the stress resultants act within the shell, a small element will be cut out, defined by two neighbouring meridians and two neighbouring parallel circles (fig. 3.3); in the limit, the element is supposed to be infinitesimal. The element can be located by the co-latitude ϕ and by a second co-ordinate θ, the longitude (fig. 3.4).

Two stress resultants are shown acting on the cut edges of the element

Domes

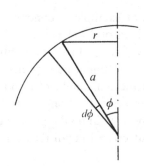

Fig. 3.2. Meridian of a spherical shell.

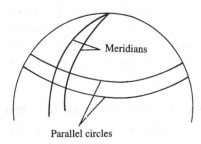

Meridians

Parallel circles

Fig. 3.3. Meridians and parallels defining an element of the shell.

in fig. 3.4; these are necessary to achieve equilibrium of the element which is subjected to its self-weight (w per unit area). The stress resultant N_ϕ acts in the meridional direction; it is shown as a tensile force in the figure, but it is to be expected that its value will turn out to be negative (i.e. compressive), and that it will increase numerically from the crown ($\phi = 0$) down to the base of the hemisphere (as is indeed the case). A second stress resultant N_θ is also shown in fig. 3.4, acting in the 'hoop' direction along the parallel circles. (In general a third stress resultant would act on the cut edges of the element, namely an action representing shear. For this particular problem of axisymmetric loading, however, the shear stress resultants are everywhere zero.)

There are thus two unknown stress resultants N_θ and N_ϕ, and it turns out that just two equations can be written, from which they may be determined, by the resolution of forces. The first of these is a differential equation:

$$\frac{d}{d\phi}\left(N_\phi \sin \phi\right) - N_\theta \cos \phi = -wa \sin^2 \phi \,, \tag{3.5}$$

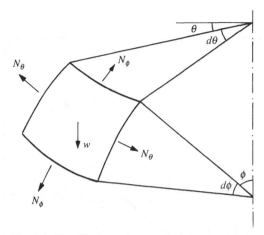

Fig. 3.4. Equilibrium of a small element of shell.

while the second is the ordinary equation

$$N_\phi + N_\theta = -wa \cos \phi. \tag{3.6}$$

The solution of these equations is straightforward, leading to

$$N_\phi = -\frac{wa}{1 + \cos \phi}; \tag{3.7}$$

the value of N_θ may then be found from equation (3.6).

The values of the two stress resultants are shown pictorially in fig. 3.5. It will be seen that, as expected, the meridional stress resultant N_ϕ is compressive throughout, increasing numerically from the value $\frac{1}{2}wa$ at the crown to wa at the support (confirming the value of equation (3.3)). However the hoop stress resultant N_θ is compressive from the crown down to a co-latitude of $51.82°$ (which is the solution of the equation $\cos^2 \phi + \cos \phi = 1$), and then becomes tensile, increasing rapidly in value towards the base.

It must be emphasized that in this analysis of the truly hemispherical shell, the stress resultants of equations (3.6) and (3.7) are determined uniquely. If the hemisphere were sliced in half (fig. 3.6), then the forces shown acting on the cut edge are necessary to maintain equilibrium. However, the tensile forces towards the base are inadmissible for the no-tension masonry structure, and clearly a different analysis must be used to explain satisfactorily the action of the masonry dome. It may be noted that the steel reinforcement in a modern reinforced-concrete shell can be designed to resist tensile stress resultants; equally, for the masonry

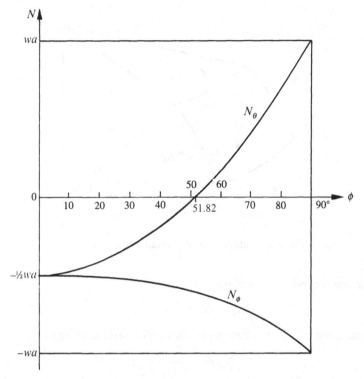

Fig. 3.5. Values of the stress resultants in a hemisphere from the crown ($\phi = 0$) to the base ($\phi = 90°$).

dome proper, it is possible to assign dimensions to rings of encircling ties which, loosely speaking, will prevent the bursting of the dome between the co-latitudes of 51.82° and 90°.

However, although the stress resultants of equations (3.6) and (3.7) are determined uniquely, they have in fact been determined from the analysis of a truly hemispherical shell of vanishingly small thickness. The actual thickness of a real dome (which can in practice be quite small) allows the mathematics to escape from this geometrical straightjacket, just as the line of thrust in the two-dimensional arch, fig. 2.5, was found not to be compelled to follow the centre line of that arch. Pursuing the analogy of fig. 2.5(a), it was seen that the inclined thrust line at the base of the arch implied a thrust of the arch against its abutments; similarly, it is to be expected that the dome will thrust out at its supports, and that these will, like the river bank, give way slightly (unless indeed the base of the dome is encircled by a tie to resist tension). Examination of the pathology of

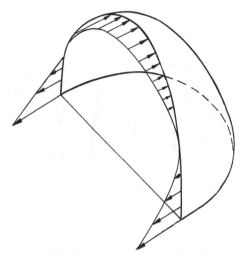

Fig. 3.6. Hoop stress resultants necessary for the equilibrium of a hemispherical shell.

the masonry dome resting on a spreading support gives an indication of the way in which the investigation of the mechanics of domes should proceed.

3.4 Cracking of domes

If it is in fact possible to find purely compressive force systems within a hemispherical dome of small but finite thickness, then it is pertinent to enquire how the dome would react to a slight yielding of its supports. A mixture of geometrical and statical analysis leads quickly to the state sketched (with great exaggeration) in fig. 3.7. From the crown down to some reasonable value of the co-latitude (which depends on the thickness ratio of the dome, but which might typically be about 25°), the dome is intact; thereafter it separates into 'slices' divided by meridional cracks. In the schematic mode of action the slices might be thought of as supporting the undamaged cap, acting somewhat as flying buttresses to the central portion.

The exaggerated pathology of fig. 3.7 is reflected in the fact that many domes show precisely such meridional crack patterns. A celebrated account was given in 1748 by Poleni, who reported on the cracks apparent in the dome of St Peter's, Rome, nearly 200 years after its completion. Poleni starts his report by making a comprehensive review of the existing

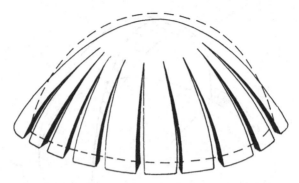

Fig. 3.7. Greatly exaggerated schematic illustration of cracking of a dome due to increase of span.

state of knowledge of masonry construction, and his FIG.XI (in fig. 3.8 here) shows his awareness of the forces necessary to maintain a critical state of equilibrium, while FIG.XII illustrates Hooke's hanging chain and the corresponding inverted arch. Further, he is aware of the theoretical and experimental French work earlier in the eighteenth century, and he states clearly and explicitly the condition for a masonry arch to be stable: that the line of thrust should lie everywhere within the masonry ('che dentro alla solidità della volta la nostra catenaria [i.e. the inverted chain] tutta intiera sia situata'). This statement is, of course, an exact formulation of the 'safe theorem' of Chapter 2.

Using these ideas, Poleni demonstrated that the dome of St Peter's was safe in its cracked state. He observes that the cracks had already divided the dome into portions approximating half spherical lunes (orange slices); for the purposes of his analysis, he hypothetically sliced the dome into 50 such lunes, one of which is shown schematically in the right-hand of the two half arches of FIG.XIII (fig. 3.8). He then proceeds to consider the equilibrium of a complete quasi two-dimensional arch formed by such a lune and its reflection.

Poleni's basic objective was to establish that a thrust line could be found for the arch which lies wholly within the material of the arch. This would then demonstrate that the dome divided into slices would be safe, and that, *a fortiori*, so would the complete dome, cracked or not. He made this demonstration experimentally. From a drawing of the cross-section of the dome (fig. 3.9), he computed the weight of the sliced arch; it will be seen that, for this purpose, each half arch was divided into 16 sections. He then loaded a flexible string with 32 unequal weights,

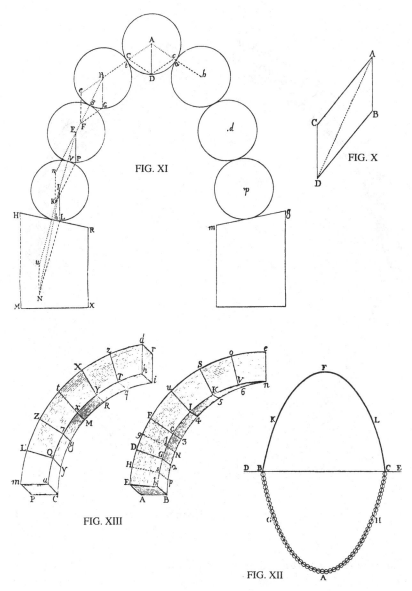

FIG. XI

FIG. X

FIG. XIII

FIG. XII

Fig. 3.8. (from Poleni 1748).

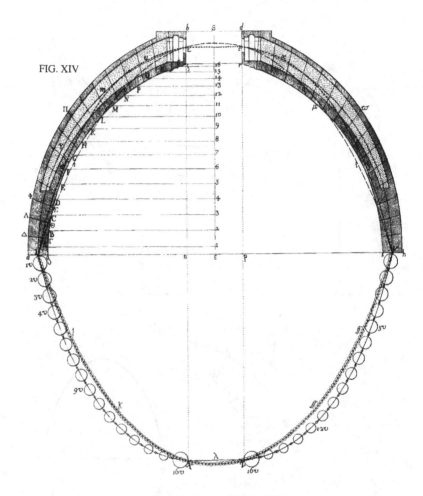

Fig. 3.9. (from Poleni 1748).

each weight in proportion to the corresponding section of the arch, with
due allowance being made for the weight of the lantern surmounting
the eye of the dome. On inversion it will be seen that the shape of the
chain does indeed lie within the inner and outer surfaces of the arch; the
other lines shown on the drawing represent the centre line of the arch,
and an inverted catenary corresponding to a uniformly loaded chain,
passing just outside the intrados in Poleni's drawing (although with a
little adjustment it can be made to fit within the masonry).

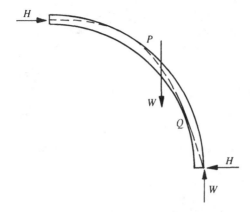

Fig. 3.10. Minimum thickness of 'orange-slice' arch.

Poleni concluded that the observed cracking was not critical, but he agreed with an earlier recommendation that further encircling ties should be provided; the inclination of the hanging chain at its points of support shows at once that there is a horizontal thrust in the sliced arch which must somehow be contained.

It will be seen from fig. 3.9 that the dome is solid only for about 20° from the springing; thereafter the dome splits into two shells between which one may climb to the lantern. In reaching his conclusions, Poleni was not disturbed by the fact that his experimentally derived thrust line fell partly in the void between the shells, and this is an instance of the deep understanding shown by his analysis. Indeed, Poleni's attack is one that would be used today.

3.5 The minimum thickness of a hemispherical dome

It was seen in Chapter 2 (fig. 2.5) that a circular arch is required to have a certain minimum thickness if it is just to contain the thrust line arising from its own weight. It is of interest to make the corresponding analysis for a hemispherical dome of uniform thickness.

As will be apparent from the final solution to this problem, it is correct to adopt the 'orange-slice' technique in which two half lunes together form a quasi two-dimensional arch. The analysis of this arch is straightforward, if somewhat complex, and fig. 3.10 indicates, approximately to scale, the correct limiting position for the thrust line. The thrust line touches the extrados at P and the intrados at Q, and passes through the extrados

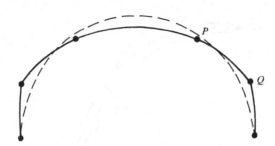

Fig. 3.11. Collapse of 'orange-slice' arch of minimum thickness.

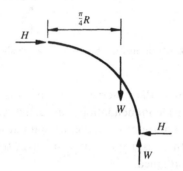

Fig. 3.12. Forces on segment of dome.

at the base of the arch. Figure 3.11 shows the corresponding collapse mechanism, for which a central extensive cap of the dome does not deform, but merely descends vertically; adjacent lunes will move apart between P and the base as the dome collapses, expressing the condition of zero tension in the masonry.

The idealized and artificial problem illustrated in figs. 3.10 and 3.11 may be explored further to give some idea of the forces necessary to maintain equilibrium of a real dome. The line of thrust in fig. 3.10 may well be imagined to approximate in shape the line of thrust determined by Poleni for St Peter's (fig. 3.9). The forces acting on the orange slice of fig. 3.10 have been redrawn in fig. 3.12; the line of action of the weight W of the segment is as shown, and the simple statics of equilibrium leads at once to the equation $H = \left(1 - \frac{\pi}{4}\right) W = 0.215W$.

Thus a hemispherical dome spanning 40 m and weighing say 150 000 kN requires support from a total horizontal force H, uniformly distributed round the base, of about 32 250 kN, that is, some 250 kN/m. If this force were to be provided entirely by a set of circumferential ties at the base, the total tie force would be 5000 kN. These figures are of the right

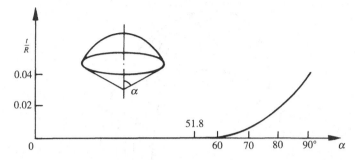

Fig. 3.13. Required minimum thicknesses for incomplete domes.

order of magnitude for the dome of St Peter's, and Poleni's agreement that extra ties should be provided was directed to the containment of the thrust. The problem is serious for such a large heavy dome, and, in the absence of proper ties, the masonry abutments to the dome must provide the necessary stability for the whole structure.

The minimum thickness of the hemispherical dome (fig. 3.10) is 4.2 per cent of the radius, that is, $t/R = 0.042$. This required minimum thickness falls sharply if the dome does not embrace a full 180° (the central dome of Hagia Sofia, for example, embraces about 140°). Thus the lune sketched in fig. 3.10 embraces a full 90° from crown to abutment; if the angle of embrace (α) is less than this, then a similar analysis leads to the required thicknesses shown in the graph of fig. 3.13. At 51.82° the thickness falls to zero; the resulting cap is in theory stable under purely compressive stresses, cf. fig. 3.6. The main dome of Hagia Sofia may therefore be expected to show little sign of those meridional cracks exhibited by domes approximating to full hemispheres.

3.6 Incomplete hemispheres

The crown of a dome can, of course, be completely omitted (i.e. the dome can be open to the sky) without any apparent structural alteration being necessary to the designs. This contrasts sharply with the action of the arch, where the removal of any voussoir, keystone or not, would break the 'chain' and lead to immediate collapse. The stresses in a dome, however (or in any vaulted structure) are distributed in two directions (fig. 3.4). This has profound implications for construction, since a masonry dome can be built almost without centering; once a circular course of masonry has been completed, it will be stable without support. (In Florence in

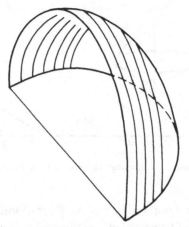

Fig. 3.14. Half dome sliced into parallel arches.

1407 Brunelleschi was not believed when he stated that he could construct a dome without falsework. He was still not believed in 1418, but was nevertheless entrusted with the work of building the segmental dome of Santa Maria del Fiore.)

A different sort of incomplete dome was illustrated in fig. 3.6. The membrane equations (3.5) and (3.6) admit only of a symmetrical solution, and would be valid for the half dome only if the free edge were subjected to the compressive and tensile forces shown. No simple membrane solution exists for the free-standing half dome. Nevertheless it is clear intuitively that a half dome *will* stand if it has sufficient thickness, and a safe estimate of this thickness may be made by using again the technique of slicing the dome into sections. Figure 3.14 shows the half dome sliced into a series of parallel semicircular arches, each arch being independent of its neighbours; the safe theorem, once again, confirms that, if the sliced half dome stands, then so will the half dome itself. The problem, then, is to determine the minimum thickness necessary for the stability of a semicircular arch under its own weight.

The solution of this problem was quoted in Chapter 2 (fig. 2.5); the thickness of the arch should be just over 10 per cent of the radius. The analysis may be made for 'incomplete' arches which embrace less than 180° (cf. the analysis for the full dome, fig. 3.13); the results for these arches are shown in fig. 3.15. For example, if the arch embraced 140° (Hagia Sofia), fig. 3.15 indicates that a half dome having a thickness of about 4 per cent of the radius should be stable.

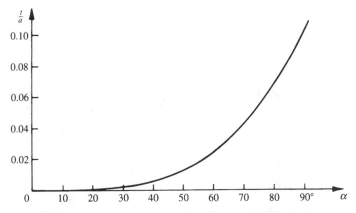

Fig. 3.15. Minimum thicknesses for incomplete arches.

Figure 3.16 reproduces Choisy's isometric drawing of Hagia Sofia; the main dome spans some 32 m, and is buttressed to the east and west by half domes, and to the north and south by massive arches capable of absorbing lateral thrust. This arrangement represents one of the three main Byzantine ways of providing support for high domes; fig. 3.17, also after Choisy, shows (a) arches on all four sides of the main dome, (b) half domes on all four sides, and (c) the arrangement at Constantinople.

Figure 3.18 may be used to explain how a half dome can act as a buttress. A complete dome has been cut into two equal free-standing half domes, and one of the half domes has been further divided into lunes. Each of these lunes will be stable if a propping force *H*, as shown in fig. 3.10, is applied at the crown. These propping forces summed for all the slices will have an out-of-balance component (of about 7 per cent of the weight of the complete dome) which acts on the undivided half dome. The half dome in fact can stand freely, and will also be stable when subjected to the out-of-balance horizontal force from the lunes. It is this capacity to resist horizontal thrusts, in general distributed round the edge, that makes it possible to use a half dome as a buttressing element.

As a structural curiosity, a three-quarter dome is sketched in fig. 3.19. One quarter of the dome of fig. 3.18 has been removed, and the remaining lunes thrust against the half dome with a force of about 2 per cent of the weight of the complete dome. This is of course less than the thrust arising from the arrangement of fig. 3.18, and it may be concluded that the three-quarter dome of fig. 3.19 is also stable. If the thickness of

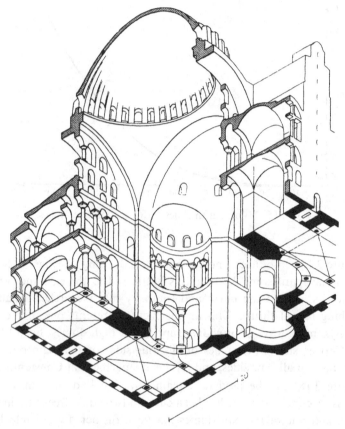

Fig. 3.16. Choisy's isometric drawing of Hagia Sofia, Constantinople (from Choisy 1883).

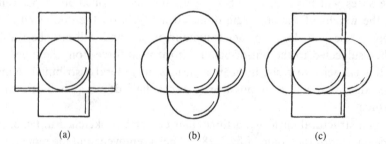

(a)   (b)   (c)

Fig. 3.17. Byzantine buttressing for a dome (after Choisy 1883).

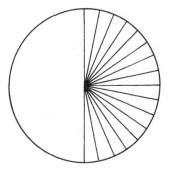

Fig. 3.18. Two half domes, one sliced into lunes.

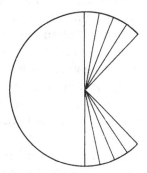

Fig. 3.19. Free-standing three-quarter dome.

masonry is such that a half dome will stand freely, then a three-quarter dome will also stand.

The main dome of Hagia Sofia has twice been in such a state. A severe earthquake of 986 led to the collapse of the western buttressing half dome, together with one-quarter of the main dome, but the remaining three-quarter dome stood. The structure was repaired, and a second earthquake in 1346 caused collapse of the eastern buttressing half dome together with the corresponding quarter of the main dome.

The difficulties in the exact analysis of incomplete domes lie in their asymmetry. A conventional membrane approach concentrates on equations written for the vanishingly thin central surface of the shell, and it is in general impossible to find solutions of these equations which also satisfy the boundary conditions (e.g. for the half dome the 'free' edge *must* be supported as shown in fig. 3.6). On the other hand a safe analysis requires the construction of a three-dimensional thrust surface lying within the masonry, and there is a better prospect of satisfying the boundary conditions in this way; however, the geometry of such a thrust surface may be forbiddingly complex.

The use of two-dimensional slices can lead to much simpler solutions; although these solutions are safe, they may so oversimplify the problem as to give too conservative results. For the types of structural shell considered in this chapter, which have t/R ratios of 5 per cent or more, the safe technique of slicing is perfectly adequate, but the technique may be too 'blunt' to deal with other, thinner shells.

3.7 Segmental domes

It was noted that Brunelleschi's dome at Florence was octangular at every horizontal section. The dome has, as at St Peter's, two shells; the total thickness of 4.2 m is composed of an outer dome of thickness 0.7 to 0.9 m, and an inner dome of 2.0 to 2.2 m. As usual, stairs in the void (of a metre to a metre and a half) give access to the lantern.

The behaviour of octagonal structures will be examined more deeply in Chapter 7, which is devoted to masonry spires. It may be noted here, however, that a circle may be circumscribed and inscribed by octagons differing very little in size. Within the thick octangular dome at Florence may be imagined a smooth dome having a circular plan at every horizontal section, and an analysis may proceed safely by using this imaginary dome as a possible thrust surface for the real structure. Good estimates may be made of the overall behaviour, and the thrust at the supporting drum may be calculated with some accuracy.

However, this blunt approach disguises an important aspect of the real behaviour, which is affected by the segmental nature of the dome; the surface does not turn smoothly, but has sharp 'creases' at the eight angles. It will be seen in Chapter 4 that such creases (groins or ribs in vaults) are potentially subject to high stresses, and may have to be reinforced. Although the shell of Brunelleschi's dome is relatively much thicker than the shell of say a quadripartite vault, nevertheless the eight ribs of the dome will tend to 'collect' the weight of the dome, concentrating the load to some extent on the corners of the octagonal drum, and so through the supporting piers to the ground.

The supports have, inevitably, given way, and the dimensions of the dome at its octagonal base have increased slightly. The resultant crack pattern is sketched in fig. 3.20. The main dome is buttressed to the north, south and east by domed apses, and on the west by the nave of the cathedral, and the cracks indicate that the north buttressing system has

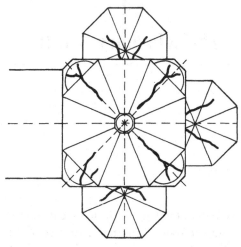

Fig. 3.20. Cracking in the dome of Santa Maria del Fiore, Florence (from Blasi and Ceccotti 1984).

moved to the north, the east to the east, and the south to the south, with perhaps the west side remaining reasonably firm. The dome has cracked through four of its shells (NE, SE, SW, NW) to accommodate these movements.

By contrast with Brunelleschi's dome, the sixteen ribs of Michelangelo's dome of St Peter's are applied to a smoothly turning shell of revolution; the ribs do not collect forces, and are not structurally effective, although they define visually the surface of the dome. A parallel will be noted in the behaviour of ribs applied to smooth vaults.

4

The Masonry Vault

The double roof system of the typical Gothic great church – a stone vault surmounted by a timber roof – is both decorative and functional. The steep external roof provides the necessary weather proofing dictated by northern climates (shallow pitches were used for Greek temples); indeed the stone vault, perhaps cracked and in any case not waterproof, itself needs the protection of the outer roof (in Cyprus the Crusader churches hardly need this cover). However timber burns well, and one function of the stone vault is to provide a fire-resistant barrier between the outer roof and the church. There is thus a symbiotic action between the two coverings of the church; the timber roof protects both the stone vault and the church from the weather, and the stone vault protects the church from the potential fire hazard of the timber roof.

The stone vault, functionally installed in a great church, at once became integrated as architectural element of the internal decorative scheme. Thus the simple quadripartite vault, lierne and tierceron vaults, net and star vaults, and the fan vault, were all developed as 'solutions' to the vaulting problem. As will be seen, these vaults, so different from each other visually, have structural actions very much in common.

4.1 The barrel vault

Figure 4.1(a) shows a straightforward 'voussoir' construction of a barrel vault – the resulting tunnel can clearly be extended to any desired length. This type of construction will need temporary formwork to support the masonry. Figure 4.1(b), on the other hand, shows an alternative design of circular vault, which can be built from tiles without the use of centering; this method was developed early in Egypt and Assyria, and can be found in some Byzantine work. The tiles are laid back at an angle (starting

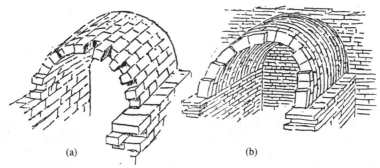

(a)   (b)

Fig. 4.1. Construction of barrel vaults (a) with horizontal courses, and (b) with inclined courses (from Jackson 1920).

from a vertical end wall), and each course acts as permanent formwork for the next course to be laid.

If each of the two vaults in fig. 4.1 is plastered after completion there will be no visual indication of their constructional difference. In discussing the action of such vaults, it is important to distinguish between what is known or imagined of their construction and what may be their actual structural behaviour. In broad terms the structural action of the two vaults of fig. 4.1 is in fact the same; each is effectively a semicircular arch, and will behave in the way discussed in Chapter 2. This is not to say that the actual structural state of the two vaults will be the same at any given time; indeed, from the discusion of Chapter 1, it is not possible to make meaningful statements about 'actual' structural states. However, a force system (Hooke's inverted chain) found to be satisfactory for one of the arches of fig. 4.1 is equally satisfactory for the other; moreover numerical calculations (e.g. to estimate the horizontal thrust of the vault on its supporting walls) will be identical for the two arches. Within more or less narrow limits each vault will impose the same thrust on its abutments.

If those abutments should spread under the action of the thrust, then it is to be expected that both vaults will develop hinging cracks roughly along their crowns, as for the semicircular arch of fig. 2.1. Detailed examination of those cracks may finally reveal something of the constructional differences of the vaults. While both cracks will run in general along the crowns, they will make small excursions as they penetrate perhaps weak mortar rather than the stronger stone. Both vaults, however, if of the same thickness and the same density of stone,

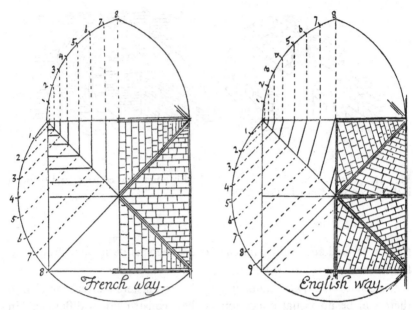

Fig. 4.2. Constructional differences in webs of Gothic cross-vaults (from Jackson 1915).

will be delivering the same thrust at their abutments, and both will have the same geometrical factor of safety that was discussed in Chapter 2.

Similarly the webs of cross-vaults (to be discussed below) may be coursed in the 'French' or the 'English' way (fig. 4.2), but the completed 'shell' structure has a basic action independent of its construction, and, once again, that construction may in fact be disguised by plaster and even by painted lines representing the stone coursing.

As a further example, the Romans constructed barrel vaults using brick tiles and mortar to produce a 'monolithic' concrete vault. Such a vault, when first completed, may be imagined to be capable of accepting tensile stresses, and thus be able to stand on its abutments without horizontal support. Movement apart of the supports will, however, crack the vault (just as the concrete of the Roman Pantheon was cracked, in a way similar to the masonry of the dome of St Peter's); the concrete barrel vault will once again have been transformed into the thrusting arch of fig. 2.1.

The barrel vault may be thought of, then, as a series of parallel arch rings, delivering a distributed line of thrust along the supporting walls. These walls must be sufficiently massive to provide the required abutment to the vault, and early medieval builders were very reluctant to pierce the walls with windows sufficiently large to give adequate illumination to the church. A first step, although very small, towards the Gothic dissolution of the wall lay in the provision of transverse arches to the barrel – a local thickening of the vault articulated visually with arches and piers of the nave arcades below. These vaulting arches are really constructional in nature; they can be built first, and act as permanent formwork for the barrels, which can then be erected bay by bay on a temporary movable scaffold. However, the arches do not concentrate the vaulting forces to any great extent; the arches will carry a small proportion of the load, but the barrel continues to act very much as before.

Further, the semicircular arch requires a massive thickness for stability, as was noted in Chapter 2. At just over 10 per cent of the radius as a minimum, a semicircular barrel vault to cover the 14 m nave of Amiens would need a thickness in excess of 700 mm. In practice the vaults need not be fully semicircular, and thicknesses may be reduced markedly on this account; fig. 3.15. Moreover, there is the possibility of rubble fill of the haunches, serving the same function as the rubble fill in the vaulting conoids of Gothic, to be discussed below; this adds further to the stability of the construction.

However, the major step in reducing structural weight comes from the use of intersecting barrels, to create the three-dimensional vault proper rather than a mere repetition of parallel two-dimensional arches. The Romans had realized the possibility of constructing two barrels intersecting at right angles to form the groin vault; each bay of the Basilica of Maxentius, in concrete (AD 30), spans over 25 m. Such a vault needs support only at the corners, and the way is at once open to introduce windows into the side walls. Further, much smaller thicknesses of vaulting material could eventually be used for these intersecting vaults, thus reducing weights on piers and lateral thrusts on external buttresses.

4.2 The cross-vault

The simplest form of groin vault results from the intersection of two equal semicylindrical barrels; the resulting bay of vaulting is square in plan, and the diagonals of the square define the location of the groins. There are geometrical difficulties inherent in the cutting of stones (stereotomy) meeting at the groins, as may be appreciated from either of the plan views of Gothic vaults shown in fig. 4.2. There are further severe difficulties if the two barrels intersecting to form the vault have

The Masonry Vault

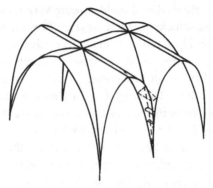

Fig. 4.3. The 'shell' of the quadripartite rib vault (after Fitchen 1961).

different spans, so that the bay is rectangular rather than square; circular barrels of different spans will of course also have different heights, which is a yet further complication. A first simplification was introduced by allowing the webs of the vault (the severies between the groins) to be domed; the groins were then not fixed uniquely by the intersection of two given semicircular barrels, but could be designed to some extent by the architect. The vaulting webs could then in turn be fashioned to fit the now fixed boundary arcs, that is, those of the groins and those at the four edges of the bay.

The groins themselves, however, were still difficult to cut, and Romanesque builders finally started their construction of groin vaults by first erecting masonry arches on the diagonals of the bay. These arches were then embedded, either wholly or partially, within the masonry of the vault webs; thus the groins were cut independently, and the rubble masonry webs erected to match. It was, of course, only a short step to build the groin arches as independent ribs, and to construct the vault webs on the backs of these. The meeting edges of the vault severies could then be cut without much care, since irregular joins could be filled with mortar and would be hidden from view by the rib.

The simplest form of the masonry rib vault is the quadripartite pattern given by the intersection of two (not necessarily circular) barrels; fig. 4.3 shows the central surfaces of these barrels as they might be modelled by an engineer wishing to use modern shell theory for the analysis of such a vault. Results from the application of shell theory are given later in this chapter; the theory may be used on the skeletal structure of fig. 4.3 to give primary vault forces, but, just as for the dome, the theory must not be pushed too far. In its simple form it is indifferent as to whether forces are tensile or compressive, whereas tensile forces cannot develop in the real vault. However, main conclusions are certainly valid – for example, that, as for the dome, the stresses in a smoothly curving vault are of the order $R\rho$, where R is the local radius of curvature of the shell and ρ is the unit weight of material. (It may be noted, again as for the dome, that doubling the thickness of the vault will double both gravity forces and the resisting area – the stress in the vault is independent of its thickness.)

A doubly curved (domed) shell has two radii of curvature (see, for example, the discussion of the fan vault later in this chapter), and strictly both values are needed in order to determine the stresses. For order-of-magnitude calculations, however, a single value suffices. As an example, a pointed vault over a nave of 15 m span might have its main radius of curvature R about 10 m; taking the unit weight of a medium sandstone as $\rho = 20$ kN/m^3, the product $R\rho$ becomes 0.2 N/mm^2, to be compared with a typical crushing stress of 40 N/mm^2.

The values of the ambient stresses in vaulting shells are confirmed as being so low that the strength of the stone is of little relevance; the level is little more than that of the 'background' compressive stress necessary to lock the stones together by friction. Thus vault webs could be constructed of light stone (as weak tufa at Canterbury), and thick mortar joints of poor (but still entirely adequate) strength could be used.

These remarks apply to smoothly curving shells. At shell intersections, as at the groins of a vault, there will be large stress concentrations. A crease in a shell introduces a discontinuity in the force field; the smoothly changing stresses have to change direction suddenly at a crease, and this cannot be achieved without the generation of large forces. The discontinuity is a line of weakness in the vault, and should be reinforced; and indeed, reinforced creases confer rigidity on the whole vaulting structure. There is, however, conflicting evidence as to the role of the 'reinforcing' ribs at the groins of a Gothic vault; the ribs at Reims seem very small, and there is the example of Longpont, Aisne, where fallen ribs nevertheless allowed the vaulting webs to stand. Once again some typical calculations furnish evidence to help resolve this conflict.

Continuing the numerical example of a quadripartite vault spanning a nave of 15 m, with a bay size of say 7.5 m, a typical thickness of the webs might be 200 mm; the webs themselves will be stressed to about 0.2 N/mm^2. The whole bay of vaulting, with its ribs, will have a weight of about 800 kN, and it is evident that the loads in each of the four diagonal ribs, at sections near the supports, will be about 200 kN. Now a rib of say 250 mm × 200 mm has an area of 50 000 mm^2, and it would be stressed to 4 N/mm^2, or to one-tenth of the crushing strength of 40 N/mm^2 – this would seem to be a safe level.

A rib of dimensions 250 mm × 200 mm could be thought of as contained within the 200 mm thickness of the main vault severies. If then the vault of this example were built without ribs, or were built on ribs that subsequently fell, the analysis indicates that the primary shell stress of about 0.2 N/mm^2 would increase sharply in the neighbourhood of the groins to the twentyfold value of 4 N/mm^2. There will be a high stress concentration, but it may be that the diagonal intersections of the severies are sufficiently regular, and the mortar sufficiently strong, that collapse of the vault does not occur. The vault will have succeeded, in a sense, in an attempt to construct its own hidden ribs.

The rib, then, serves a structural purpose as a very necessary, but perhaps not finally essential, reinforcement for the groins; it enables vaulting compartments to be laid out more easily; it enables some constructional formwork to be dispensed with; and it covers ill-matching joints at the groins. As a bonus, the rib has been thought to be satisfying aesthetically, and all of these functions may be thought of as the 'function' of the rib.

The structural purpose is, however, clear – a sharp crease in a shell demands reinforcement. Thus in a quadripartite nave vault formed by intersecting barrels (pointed or not) with a level soffit, the deep diagonal creases (fig. 4.3), increasing in sharpness towards the springing of the vault, require reinforcement, and the diagonal ribs at these creases (or the 'hidden' ribs within a plain groin) emerge as the effective members carrying the whole vault. By contrast there will be no creases in the shell at the nave walls or at the positions of the transverse arches (visually dividing the nave into its bays); neither the wall arches (formerets) nor the main transverse arches are required to carry anything but their own weights.

If the soffit of the vault is not level, as when each severy is gently domed, for example, then a crease will appear at the transverse arches – these then become true structural elements, although they will be loaded only slightly if the crease is shallow. This is one example of a general rule. Ribs applied to a crease in a shell surface act as reinforcement; ribs applied to a smoothly turning surface (e.g. tiercerons and lierne ribs) are decorative.

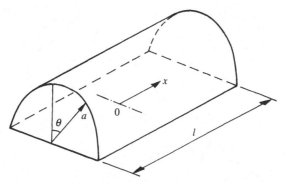

Fig. 4.4. Semicylindrical shell carrying its own weight.

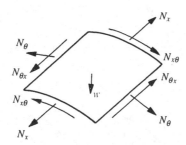

Fig. 4.5. Element of shell of barrel.

4.3 Vault thrusts

The semicircular barrel vault may be examined by membrane shell theory, as a first step in the understanding of the way in which a cross-vault carries its own weight, and distributes forces to its supports. As has been noted, care must be taken in applying shell theory to masonry, but essential features of the behaviour of the vault are revealed by this simple analysis.

Figure 4.4 gives a sketch of the barrel, of radius a; an element in the surface may be located by the distance x along the axis and the angular displacement θ from the vertical.

Such an element is shown in fig. 4.5; the weight w per unit area produces three stress resultants in the shell, labelled N_x, N_θ and $N_{x\theta}$ ($= N_{\theta x}$). The first two of these represent direct stresses of compression (or tension), while the third represents a shear stress. The full equations will not be written here, but it may be noted that equilibrium must be satisfied in three directions, so that three equations are available to determine the

Fig. 4.6. Intersecting semicylindrical shells forming bays of a cross-vault.

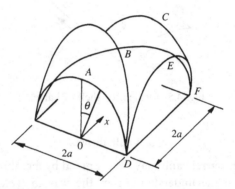

Fig. 4.7. Single bay of the vault of fig. 4.6.

three stress resultants. The equations are not completely straightforward, but one is particularly simple; resolution of forces in the radial direction gives immediately

$$N_\theta = -wa\cos\theta. \qquad (4.1)$$

Thus the compressive stress resultant round the circumference of the barrel has value wa/unit length at the crown, and decreases (by the factor $\cos\theta$) as the section under scrutiny moves from the crown.

The equations, and in particular equation (4.1), are the same for each of the vaulting severies of the shell shown in fig. 4.6. Here a continuous barrel of radius a (whose axis lies along the nave) is intersected at right

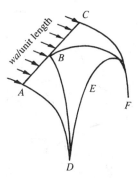

Fig. 4.8. Forces on half bay of vault.

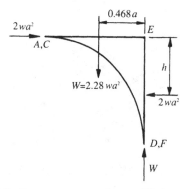

Fig. 4.9. Forces necessary for equilibrium of vault.

angles by a succession of similar cylinders, and the resulting vault serves as a crude model of the real structure. Figure 4.7 shows a single bay of this vault. Now the remains of the barrel lying along the nave are, so to speak, unaware of the intersecting barrels; the stress resultant N_θ in the portion ABD of the shell, for example, is still given by equation (4.1). Thus if the vaulting bay of fig. 4.7 is cut in half along the crown, as in fig. 4.8, then a compressive force of magnitude wa/unit length must act along the cut ABC in order to maintain equilibrium.

The total sideways thrust of $2\,wa^2$ (which is in fact the thrust of the vault, bay by bay) must be resisted externally, as shown in the side elevation of the half vault (fig. 4.9). For this idealized example the weight W of the half vault is $2.28\,wa^2$, and this must be resisted by an equal and opposite vertical propping force at the springing. Figure 4.9 shows, then, the resultant set of forces keeping the vault in equilibrium, where

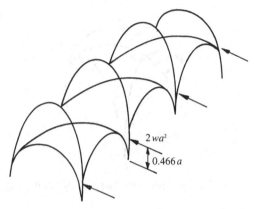

Fig. 4.10. Location of flying buttresses to counteract the vault thrusts.

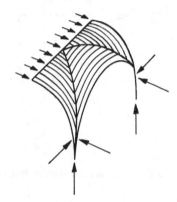

Fig. 4.11. Vault sliced into parallel arch rings.

the weight W acts through a known position (0.468 a from the edge of the vault). These forces must also achieve a moment balance, and this requires the external propping thrust 2 wa^2 to act at a particular distance h below the crown of the vault; simple statics gives at once $h = 0.534\ a$.

In aisled Gothic construction the horizontal propping forces to each bay are provided by the flying buttresses, and these buttresses should therefore be positioned as shown in fig. 4.10. Although the values marked in fig. 4.10 will not be those arising from the analysis of a real Gothic vault (rather than from the idealized semicircular barrels), nevertheless the opposing thrust of the flying buttresses acts at some distance above the springing of the vault, and it is essential to provide fill to the vaulting conoid, as indicated in the more realistic sketch of fig. 4.3. This may be

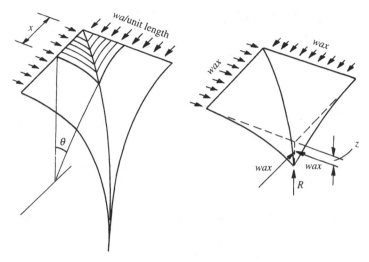

Fig. 4.12. Portion of the vault of fig. 4.11.

thought of as a stabilizing mass acting on the lower portions of the vault, but in fact the fill can be made of lightweight material; the essential function of the fill in the vaulting pocket is to provide a pathway for the vault thrust as it escapes from the diagonal ribs.

The slicing technique illustrated in the previous chapter on domes may be used again to give some further understanding of the way in which forces are distributed in the cross-vault. Figure 4.11 shows the half bay of vaulting of fig. 4.8 sliced into independent arches, not interacting with each other, and therefore supported solely from the diagonal cross-ribs reinforcing the groins. If it is supposed for the moment that the thrust in each sliced arch at its crown continues to have the previous value, wa per unit length (other values are considered below), then the forces in fig. 4.9 are unchanged, and in particular h continues to have the value $0.534\ a$. Equilibrium of a portion of the vault (fig. 4.12), requires that the balancing thrust 'in' the rib in fact acts above the rib, at a distance marked z in the figure. An analytical expression may be written for z, and the results are shown graphically in fig. 4.13, where the path of the thrust is shown by the broken line. It will be seen that this thrust line gradually parts company from the centre line of the rib. For some considerable distance from the crown a rib of finite thickness will be able to accommodate the thrust, but eventually it will pass into the fill in the vaulting pocket.

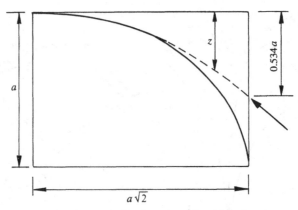

Fig. 4.13. True view of diagonal rib of vault of fig. 4.11, showing path of thrust 'in' the rib.

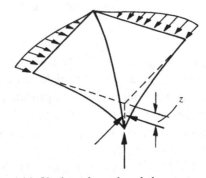

Fig. 4.14. Vault under reduced thrust at crown.

The assumption just made of a uniform crown thrust of magnitude wa per unit length is not very realistic, even for this idealized example. A very thin circular arch of small angle of embrace will necessarily have a thrust approaching this value, but an arch of finite thickness will be stable under a much smaller value of thrust. For a calculation which explores the action of the vault under these smaller thrusts, the sliced arches of fig. 4.11 will be taken to have thicknesses $t/a = 0.02$ (e.g. a 150 mm vault over a span of 15 m); each of these arches will be supposed to be carrying its minimum possible thrust (as illustrated, for example, in fig. 2.2(a)). Thus for a portion of the vault (fig. 4.14), the edge loading at the crown will not have the uniform intensity wa but will act at lesser values, as sketched, the actual value at any section depending on the

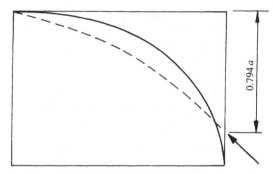

Fig. 4.15. Path of thrust in the diagonal rib of the vault of fig. 4.14.

Fig. 4.16. Sliced pointed quadripartite vault.

distance away from the crown. The calculations may be carried through as before, and the resulting thrust line in the rib is sketched in fig. 4.15.

The actual value of the horizontal thrust has dropped from $2\ wa^2$ to $1.35\ wa^2$, so that the flying buttress should be placed lower than before. However, the significant feature of fig. 4.15 is the immediate departure of the thrust line from the diagonal rib at the crown; this almost compels the adoption of the pointed Gothic form rather than the circular Romanesque. A flatter pointed diagonal rib is better suited to the thrust line of fig. 4.15 than the elliptical form shown, in the sense that the rib centre line may be made to coincide more nearly with that thrust line.

The calculations were made by considering circular arches, but there is no difficulty in making a similar analysis for a pointed vault, as in fig. 4.16. As before, the shell of the vault may be sliced into parallel

Fig. 4.17. Minimum thrust in a pointed arch.

arches, not interacting with each other, and supported by the diagonal ribs below the shell intersections. Lines of thrust may be drawn for each individual arch slice, and corresponding supporting forces determined. Figure 4.17 illustrates the condition of minimum supporting thrust for a particular arch segment, and it will be seen that the forces cannot 'reach' the tip of the extrados of the pointed arch; this behaviour is of some significance in the interpretation of the pathology of Gothic vaults, discussed below.

It should be emphasized again that the pattern of forces resulting from this kind of analysis is safe, even though the actual structure may experience a different distribution. Moreover, essential design parameters may be determined with some confidence – the weight of the vault can be calculated accurately and good estimates may be made of the supporting forces (from the flying buttresses or other abutments).

4.4 Ungewitter's tables

It will be appreciated that the calculations given above refer to an idealized shell of uniform thickness, with no allowance made for the weights of the ribs, bosses, or vault fill. These quantities may be estimated for any given existing vault, and similar calculations will then lead to values of the quantities of interest – in particular, of the vault thrust and its line of action. One calculation is very much like another, with the variable parameters being the length and breadth of the vaulting bay and its rise from springing to crown, and the thickness and material of the vault webs.

Such calculations were made in a 'smoothed' form and tabulated by Ungewitter (1901). The rise to span ratio of the vault is one important variable, and in table 4.1 (which is a reduced form of one of Ungewitter's tables), this ratio takes five values, from the very low rise of 1/8th of the span, through a rise equal to half the span to a rise of twice this value. Ungewitter quotes results for five thicknesses of vault (and gives partial

Table 4.1. *Thrusts from quadripartite vaults (after Ungewitter 1901)*
(V_o and H_o in kN/m^2)

Height/span f/s	1:8		1:3		1:2		2:3		5:6 to 1:1	
kN/m^2	V_o	H_o	V_o	H_o	V_o	H_o	V_o	H_o	V_o	H_o
a. $\frac{1}{2}$ lightweight brick	2.0	3.6–4.0	2.3	1.6–1.8	2.6	1.1–1.2	2.9	0.9–1.0	3.4	0.8–0.9
b. $\frac{1}{2}$ strong brick	2.7	5.0–5.5	3.1	2.2–2.4	3.5	1.4–1.6	3.8	1.1–1.3	4.5	1.0–1.1
c. $\frac{3}{4}$ strong brick	3.7	7.0–7.5	4.2	3.0–3.3	4.8	1.9–2.2	5.3	1.6–1.8	6.5	1.5–1.6
d. 200 mm sandstone	5.0	9.5–10.0	5.7	4.2–4.5	7.0	2.8–3.2	7.5	2.2–2.5	9.0	2.1–2.3
e. 300 mm rubble	8.5	16–17	10.0	7.1–7.5	12.0	4.8–5.5	13.0	4.0–4.3	15.0	3.5–3.7
lever arm h/f		0.90		0.85–0.75		0.80–0.70		0.80–0.72		0.80–0.75

values for a vault filled level to form a floor, as say a triforium floor over a side aisle, for which it is not possible to give a complete general tabulation).

The five cases given by Ungewitter correspond to standard brick sizes ($250 \times 120 \times 65$) of weight 16 kN/m^3 (or 12–13 kN/m^3 for lightweight bricks), and to sandstone of 20 kN/m^3 and rubble of 24 kN/m^3. The vertical weights V_o for each of the five cases are calculated for unit *plan* area of the vaults, and include allowances for the ribs, etc. The vault thrust H_o is tabulated in the same way. The five vaults considered are

a. $\frac{1}{2}$ lightweight brick (i.e. 125 mm)
b. $\frac{1}{2}$ strong brick (125 mm) or $\frac{3}{4}$ lightweight (190 mm)
c. $\frac{3}{4}$ strong brick (190 mm) or 1 lightweight (250 mm)
d. 1 strong brick (250 mm) or 200 mm sandstone
e. 300 mm rubble vault

Figure 4.18 marks the parameters on a sketch of the vault.

As an example of the use of the table, the vault thrust will be calculated for a 200 mm sandstone vault covering a nave bay 16 m × 8 m, with a rise f (cf. fig. 4.18) of 8 m. This is Ungewitter's class d, with a height to span ratio of 1:2, so that $V_o = 7.0$ kN/m^2 and $H_o = 2.8$–3.2 kN/m^2. The value of h/f is 0.70–0.80 (cf. the value 0.794 marked in fig. 4.15), and it is required to calculate the vault thrust H (to be resisted by the flying buttress) marked in figs. 4.19 and 4.20. The weight V (fig. 4.20), corresponds to that of half the complete bay of vaulting, so that

$$V = \left(\tfrac{1}{2}\right)(16)(8)\,V_o = 448 \text{ kN}$$

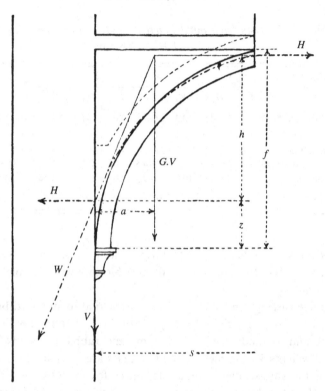

Fig. 4.18. Parameters involved in the calculation of vault thrust, table 4.1 (after Ungewitter 1901).

and similarly $H = \left(\frac{1}{2}\right)(16)(8)H_o = 179 - 205$ kN. Thus the vault thrust is determined from Ungewitter's tables as lying between about 18 and $20\frac{1}{2}$ tonnes.

Similarly the two thrusts P marked in fig. 4.19, cancelling out bay by bay but acting eventually on the westwork of the church (and on the east end of the choir), may be calculated using a height to span ratio of 1:1; the values of V_o and H_o are 9.0 and 2.1–2.3 kN/m^2 respectively. Each of the forces P looks after one quarter of the bay, so that

$$P = \left(\tfrac{1}{4}\right)(16)(9)(H_o) = 67\text{--}74 \text{ kN},$$

that is, about 7 tonnes.

The values of the horizontal thrusts H and P depend on the value of V and also on its line of action (the parameter a marked in fig. 4.20, calculated by Ungewitter but not shown in the reduced table 4.1). The

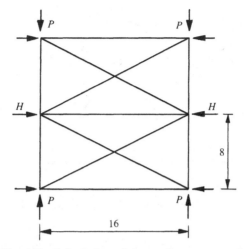

Fig. 4.19. Calculation of thrusts for a nave vault.

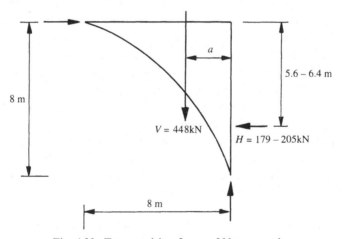

Fig. 4.20. Forces arising from a 200 mm vault.

use of the table will give the designer a ready and accurate measure of vault thrusts, but of course independent and prime calculations may be made to check the values for any particular church. Knowledge of thrusts is clearly of value if, for example, a flying buttress has to be taken down and rebuilt, since temporary shoring can be designed with some confidence.

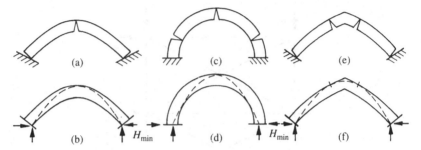

Fig. 4.21. Crack patterns in arches resulting from increased spans.

4.5 The pathology of the quadripartite vault

The structural behaviour of the voussoir arch was approached through its pathology – it was remarked that in order to accommodate an increased span, the arch would develop cracks. Figure 4.21 shows crack patterns for three arches of different shapes, and in each case three hinges have formed, making the arches statically determinate under minimum values of abutment thrust. Actually it will be seen from figs. 4.21(e) and (f) (cf. fig. 4.17), that a pointed arch should in theory form four hinges under these conditions. In fact it is evident that any slight asymmetry, whether of geometry of the arch or of loading, will ensure that only one of the hinges near the crown will form. In each of the cases (a), (c) and (e) of fig. 4.21 a crack would be visible at or near the crown when viewed from below, and cracks of this sort can often be seen.

There are three identifiably different types of crack which can occur in masonry vaults. Figure 4.22 reproduces Pol Abraham's drawing (1934) of a typical quadripartite bay, and his identification of the cracks, which all run roughly east–west, is as follows:

1 There are cracks in the main barrel of the vault in the region of the crown; these are hinging cracks corresponding to the sketches of fig. 4.21.

2 There are the cracks that Abraham called *fissures de Sabouret*, parallel to the wall ribs but some way in from the wall. As will be seen, these cracks involve complete separation of the vault webs, so that, standing on top of the vault, one may see the floor of the church below.

3 There are often similar cracks separating the vault webs from the north and south walls.

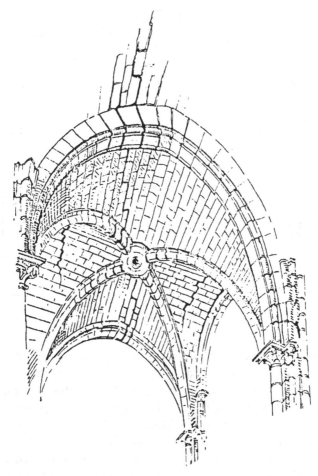

Fig. 4.22. Typical cracks in Gothic vaults. Pol Abraham (1934) distinguished between the 'hinging' cracks near the crown, the 'Sabouret' cracks parallel to the wall ribs, and the separation of the vault from the walls.

(Sabouret (1928) gave some of the taxonomy of vault cracks.)

As a first step towards the understanding of the way that these three types of crack arise, the behaviour of a uniform barrel vault may be studied. Figure 4.23(a) shows the cross-section of such a vault, drawn roughly to scale (say a vault thickness of 300 mm with a span of 12 m). The vault is supposed to be maintained by external supports. As drawn, the arch is too thin to carry its own weight right down to the springings, and rubble fill is shown backing the haunches so that the thrust can

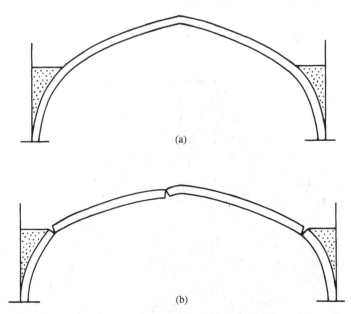

Fig. 4.23. Hinging crack in barrel vault resulting from yielding of buttress system.

escape from the vault proper. In fig. 4.23(b) the buttressing system is supposed to have given way slightly, and the familiar crack pattern of three hinges has developed. The hinges in the extrados near the fill will not be visible from below, but the more or less central crack is the first of Pol Abraham's three types of defect.

The vault of fig. 4.23 is essentially two-dimensional in the sense that the cross-section was supposed to be the same down the length of the church. Figure 4.24 shows a single bay of a quadripartite vault formed by the intersection of two slightly pointed barrels. In fig. 4.24(a) an elevation of the vault is shown, looking east down the length of the church; the fill, which serves the same function as before, is placed in the vaulting conoids. (Cf. the plan of fig. 4.24(c). The vault is of course supposed to extend for several bays.) The vault thrusts, passing through the fill, will act on the external supports, either main buttresses in aisleless construction, or flying buttresses transferring the thrusts over side aisles. If now the buttressing of the vault gives way, the portion which runs east–west will crack as before, and the single hinge line near the crown will be seen from within the church. The change in geometry necessitated by the increase in span will result in a drop of the crown of the vault.

There is, however, a severe geometrical mismatch in the intersecting

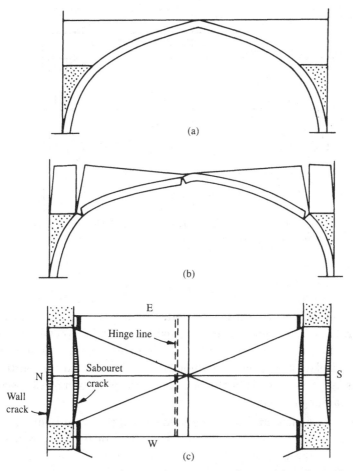

Fig. 4.24. Crack patterns in one compartment of a quadripartite vault resulting from yielding of buttress system.

barrel which runs north–south; there is not enough masonry to fill the increased north–south dimension. A crack pattern which allows the vault to deform in virtually strain-free monolithic pieces is sketched in figs. 4.24(b) and (c). Cracks (the Sabouret cracks) have opened in the vault at a distance of a metre or so from the north and south walls (usually containing windows). In addition, cracks may have opened adjacent to the walls. Thus cracks near the crown are traces of hinge lines in a portion of the vault through which compressive forces are being carried; in fact the forces act perpendicular to the hinge lines. By contrast

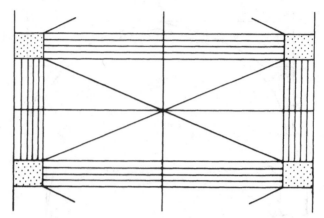

Fig. 4.25. Schematic plan of a bay of vaulting; cf. fig. 2.24(c). The edges of the vault act as simple arches.

the wall cracks and Sabouret cracks indicate complete separation of the masonry, through which a hand may often be passed. Clearly no forces can be transmitted across such fissures, and in fact the compressive forces run parallel to the cracks. A wall crack and an adjacent Sabouret crack effectively isolate a portion of the north–south barrel, and this portion then acts as a simple arch, a metre or so in width, and spanning east–west between vaulting conoids.

In principle similar cracks might develop in a direction at right angles, in the east–west barrel. However successive bays of the vault buttress each other in the east–west direction, preventing the increase in span which would lead to fissures. At the west end of the cathedral the westwork is usually sufficiently massive to maintain the last bay of the vault in position, but the east end of the church may be of less robust construction. In this case a wall crack in the main vault may be seen above the east window, again with a maximum complete separation of the fabric on the axis of the church, tapering to zero at the north and south walls.

Thus the structural action of a vaulting bay may be envisaged as in the sketch plan of fig. 4.25. The edges of the vault may be thought of as two-dimensional arches spanning between the vaulting conoids, and may in many cases have been transformed into such arches, if Sabouret cracks are visible. Whether visible or not, this is potentially the most severe condition for the vault edges, where the spans are greatest. The central quadripartite portion of the vault in fig. 4.25 has relatively much greater strength; it is the edges of the vault which are critical.

4.6 Superimposed loads on vaulting

The whole discussion above has referred to the action of a vault under its own self-weight; the loading is dead, and it is asymmetrical. Wind forces will not be transmitted to the vault; forces on the timber roof will be carried to the walls of the church, and to the ground by the buttresses. Under normal circumstances, therefore, only symmetrical dead-load conditions are experienced by a masonry vault.

However, environmental catastrophes can impose unsymmetrical loading on the roof system. One such catastrophe is an earthquake; it was noted that many ancient buildings, Greek and Roman temples for example, survived earth tremors, although there is no doubt that complete or partial collapses also occurred. Analysis of earthquake effects is beyond the scope of this book, but the subject is of importance with respect to the buildings of southern Europe, and has received some attention in recent years.

It is possible to make some remarks about the consequences of a fire in a church, although firm conclusions can be reached only for specific examples. However, numerical calculations at least indicate the relative importance of various effects. Thus a masonry vault might have a self-weight of 5 kN/m^2 (see table 4.1); the great timber trusses above the vault, together with their boarding and lead or tile covering, might weigh 3 kN/m^2. It is this extra weight which would be imposed on the masonry vault if the timber roof should collapse during a fire.

If the extra load were imposed uniformly then the vault would not be distressed. The mere increase of a load from an intensity of 5 kN/m^2 to 8 kN/m^2 will have no effect on the geometry of the thrust lines within the masonry, and there will be no danger of hinging collapse. The same conclusion is valid whatever the increase of load (at least, up to a very large value, when crushing of the stone might result), so that a dynamic collapse of the timber roof, leading to larger effective forces, could be accommodated provided still that the imposed loading were uniform.

The superimposed loading will, of course, not be uniform. Accordingly, the carrying capacity of the vault under other load distributions must be investigated. It was noted at the end of the last section that the edges of the vault were the critical regions; in fig. 4.26 the portion *ABC* of the vault adjacent to the wall may have become separated from the

The Masonry Vault

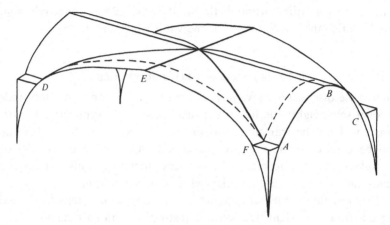

Fig. 4.26. Single bay of quadripartite vaulting.

main structure by the development of Sabouret cracks. Whether or not it is physically separated it will tend to act as a quasi two-dimensional arch, and a study may be made of the arch considered in isolation. (The calculations can be made quickly by hand, but computer programs also exist, since the arch problem is of importance in the assessment of masonry highway bridges.)

The most severe loading on the arch occurs when a point load acts at about quarter span, as shown schematically in fig. 2.3; at a sufficient intensity of the load P the self-weight of the arch will be 'overcome', and collapse will occur by the development of four hinges. Thus, in fig. 4.26, if the portion of the roof above the half-arch AB were to fall somewhere near the centre of AB, this might lead to collapse. A definite prediction could be given only for a specific vault with known overall dimensions and thickness; however, a roof weighing 3 kN/m² falling on a vault of usual construction will impose a load that is of the right magnitude to cause collapse.

Equally serious conditions might be imposed on the vault by those fighting the fire; water will give rise to extra loads if this accumulates in the vaulting pockets. If the four pockets in fig. 4.26 were full of water, for example, the loading on the whole vault would still be symmetrical, but it is likely (again for usual dimensions) that the vault would not be able to withstand the resulting fluid pressure.

An even more dangerous case would arise if the loading is unbalanced. If the quarter $EFAB$ of the vault in fig. 4.26 is filled with water, for example, then the pressure on the portion AB of the critical edge arch will not be balanced by a corresponding pressure on portion BC.

Calculations such as these do not encourage optimism about the safety of a vault in the case of fire in the timber roof above. The calculated collapse loads are of the same magnitudes as those of the loads liable to be imposed by the fall of the high roof, and dynamic effects could well lead to considerable damage. Further, punching phenomena have not been discussed; a heavy fall on a more or less localized area of the masonry will lead to a local hole in the vault. Further, the weight of water that might build up in the vaulting pockets represents a severe loading case.

On the other hand, none of the damage which may occur need necessarily be widespread. In fig. 4.26 the broken line parallel to ABC represents the position of a Sabouret crack, and isolates the weakest portion of the vault running in this direction. If a Sabouret crack is physically present, then water cannot build up in the vaulting pocket. If the cracks have not developed (or, more likely, have been repaired as a consequence of good maintenance of the church), then the water pressure could build up slowly and lead to collapse of the arch ABC, but will not necessarily lead to spreading collapse of the whole vaulting system. On the contrary, the main portion of the vault could well remain intact.

Similarly, a hole punched through the arch ABC by collapse of the high timber roof would lead to collapse of that arch, but leave the main severies of the vault untouched. A hole punched through a severy near the centre of the vault could well remain a stable hole. The masonry vault has a high intrinsic strength, and is well able to resist local damage.

4.7 The sexpartite vault

As an example of a more complex vaulting pattern (the fan vault is considered separately in the next section), the ideas of this chapter may be applied to the sexpartite vault. Figure 4.27 shows a schematic view of this vault; cf. fig. 4.3 giving the quadripartite arrangement. In the sexpartite pattern, the bays are vaulted in pairs; the ridge ribs do not run directly north–south, but connect alternate bays as shown (ribs CC in fig. 4.28).

The 'creases' in the shell of fig. 4.27 indicate, as before, those ribs which carry. Thus the main transverse (north–south) ribs AA in fig. 4.28 carry merely their own weight; the intermediate transverse ribs BB, however

The Masonry Vault

Fig. 4.27. The 'shell' of the sexpartite vault (after Fitchen 1961).

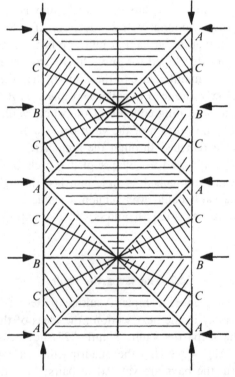

Fig. 4.28. Plan view of the sexpartite vault of fig. 4.27.

are loaded by the vaults ACB on either side. Similarly the *diagonal* ribs AA carry high loads.

Notre-Dame in Paris was vaulted in the sexpartite pattern in the 1160s, and there has been discussion (see Conant 1944) as to whether or not the original flying buttress system was applied to alternate bays only, that is, to the vaulting conoids at A in fig. 4.28 but not to those at B. Exact calculations can be made only for a particular vault with known dimensions, but the slicing technique may be used to give general indications of behaviour. If the sexpartite vault is divided into arches (similar to those for the quadripartite vault of fig. 4.16), then it may be concluded quickly that the thrust at the position of a main transverse rib A is about three times as great as that at an intermediate rib B. Thus it could well have been possible to dispense with the intermediate flying buttresses provided that the (existing) strip buttresses were adequate to take up the load.

This discussion based upon numerical calculation hardly answers the question of whether or not Notre-Dame was originally built with the intermediate flying buttresses omitted. However, it does indicate that there is a real question of architectural history to be answered, and one which cannot be dismissed summarily on technical grounds.

4.8 The fan vault

In France there was little development of the rib vault from the quadripartite and sexpartite forms, whereas complex net and star vaults were constructed in southern Germany, Austria and Bohemia, and lierne and tierceron vaults in England. Many of these decorative vaults were based on the quadripartite pattern, however; the tierceron ribs, for example, can be stripped away from the smooth vault surface to reveal the basic form.

One English development which departs radically from the standard form was that of the fan vault. The ribs of the fan are again applied to a smoothly turning surface, and their function is to define visually the shape of the vault but not to carry load; however, the shell beneath the ribs is fundamentally different from those considered so far. A shell surface has two principal curvatures – two lines drawn on the surface, and meeting at right angles, have in general two different curvatures at each point. Thus the vaulting webs of fig. 4.16 are curved in one direction (that of the sliced arches) but flat in the direction at right angles. In mathematical terms one of the principal curvatures is zero,

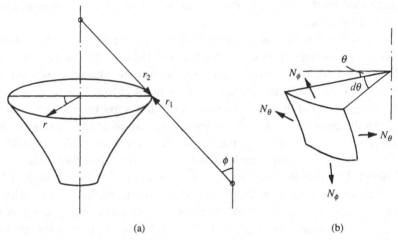

(a)   (b)

Fig. 4.29. (a) The complete shell of revolution underlying the fan vault; (b) a small element of the shell.

so that the Gaussian curvature (the product at each point of the two principal curvatures) is also zero.

A domed vaulting severy no longer has a flat surface, but has two positive principal curvatures; viewed from below, the curvatures of both lines meeting at right angles are concave, and the Gaussian curvature is positive. In the same way the domes of Chapter 3, formed as shells of revolution by rotating a curve (not necessarily a circle) about an axis, also have positive Gaussian curvature. By contrast the fan vault, shown schematically as a complete shell of revolution in fig. 4.29(a), has a negative Gaussian curvature; of two principal lines on the surface, one is concave and one convex. Just as for the dome, a fan vault of given profile can be analysed readily by membrane shell theory, and fig. 4.29(b) shows the stress resultants, acting on the edges of a small element, which arise from the loading imposed on the shell (cf. fig. 3.4). The usual straightforward equations may be solved to give the values of the hoop stress resultant N_θ and the meridional stress resultant N_ϕ for each element of the shell. Once again the theory may require tensile forces to be developed for certain shapes of fan, and such solutions would be inadmissible for a masonry vault.

Of interest, therefore, are solutions of the basic equilibrium equations for which the hoop stress resultant N_θ is zero; that is, the shape of fan is sought for which the forces are carried down the meridians only. Such

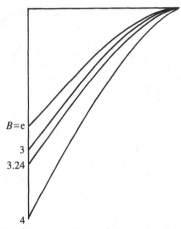

Fig. 4.30. Profiles of fan vaults carrying meridional forces only.

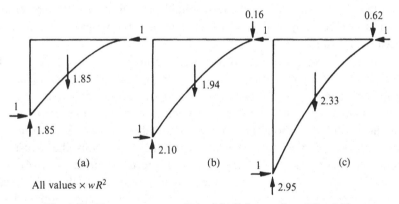

All values × wR^2

Fig. 4.31. Forces on some of the 'ideal' fan vaults of fig. 4.30.

a vault could be sliced along the meridians; in particular, the complete shell of fig. 4.29(a) could be cut into two fan vaults by a vertical plane without any forces being introduced at the cut, and forces in the vault would be carried down to the springing and so to the external buttressing system. It turns out that there is an infinite range of such profiles, of which four are sketched in fig. 4.30. It may well be imagined that one or other of these profiles could be matched to a given fan vault; the point is explored further below.

However, to continue for the moment to discuss some of the results of this theoretical analysis, fig. 4.31 shows side views of three of these ideal fans with the forces marked which are necessary to preserve equilibrium;

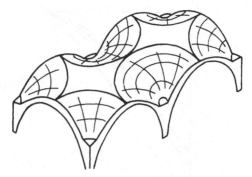

Fig. 4.32. Two bays of fan vaulting.

all the numbers shown should be multiplied by wR^2 where R is the maximum radius of the fan (i.e. half the width of the space being vaulted). The value of w is the true weight per unit area of the material, that is, it is not adjusted as in table 4.1 to give the weight per unit *plan* area of the vault.

It may be noted that the thrust at the abutment is the same for all the profiles of fig. 4.31, the value being simply wR^2. As an analogue of King's College Chapel, Cambridge, which has a span of about 12.8 m, the value of R would be 6.4 m; the thickness of the (limestone) panels in the vaulting is about 125 mm, so that, allowing for the decorative ribs, the value of w might be taken at 4 kN/m^2. Thus the thrust at King's College Chapel in each bay would be determined as $4(6.4)^2$ or about 164 kN. This is a first estimate; a deeper analysis, along the lines discussed below, leads to the conclusion that the thrust could well be less than the 16 tonnes implied by this calculation.

Secondly, it may be noted from fig. 4.31 that the top edge of the fan must in each case be subjected, somehow, to a horizontal force of wR^2, and also to a vertical force which varies according to the actual profile; these balancing forces are in fact distributed round the top edge of the fan. It will be appreciated from the sketch of fig. 4.32 that the fans which have been analysed form only a portion of the complete roof system. If the bays are square, as in fig. 4.32 and fig. 4.33(a), then the fan vaults proper will be separated by more or less domical areas of spandrel. For rectangular compartments the fans may intersect (fig. 4.33(b)), as at King's College Chapel, or they may be complete (fig. 4.33(c)), but again there will be a central region of spandrel.

Figure 4.34 shows Robert Willis's (1842) sketch of the vault to the

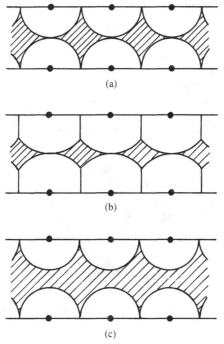

(a)

(b)

(c)

Fig. 4.33. Geometrical constraints on fan vaulting.

retrochoir at Peterborough Cathedral, of span about 8.2 m; the construction at King's College Chapel is very similar (and perhaps John
Wastell was the architect of both). Of particular notice is the rubble fill
in each vaulting conoid. Just as for the rib vault, such fill enables the
thrust surface to pass out of the fan-vault shell into more or less solid
material; even without the fill, it is evident that the fan proper stops
well short of the apparent springing. Figure 4.35 shows schematically
a vaulting compartment at Peterborough (or at King's College Chapel),
and three distinct regions may be noted. The fan itself is of limited
extent, merging into the fill at its base. On the centre line of the bay
the spandrel of shallow or zero rise bridges between the fans, and this
spandrel masonry, carrying perhaps bosses and pendants, provides the
horizontal and vertical forces marked in fig. 4.31 which are necessary to
maintain static equilibrium.

Since the fan proper is so limited, it seems clear that it should be
reasonably straightforward to fit one of the 'ideal' profiles of fig. 4.30
within the thickness of the masonry. Indeed, fig. 4.36 shows a cross-

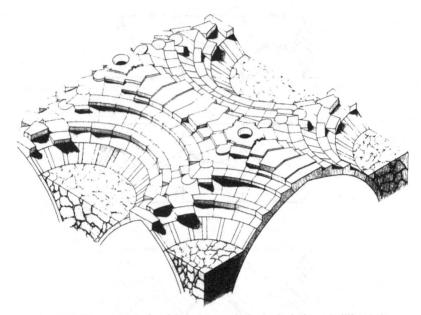

Fig. 4.34. The retrochoir of Peterborough Cathedral (from Willis 1842).

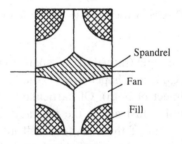

Fig. 4.35. Plan view of vaulting compartment.

section of the vault at Peterborough, and it will be seen that the 'ideal' thrust line matches the actual profile pretty well. Note that the effective radius R of the fan is less than the half-width of the vaulted space, so that the vault thrust wR^2 is correspondingly reduced; the spandrel masonry, weighing nearly $\frac{3}{4}$ tonne/m provides an effective edge loading. The whole of this analysis shows, in fact, that a fan vault is an efficient form of roof system. A late fifteenth-century masonry designer specializing in fan vaults could pick any likely-looking profile, could decorate the surface with non-structural ribs and could insert on the centre line of the nave

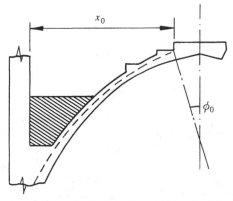

Fig. 4.36. Cross-section of a fan vault (Peterborough) with 'ideal' line of thrust.

a series of heavy bosses, all with the assurance that his structure would be stable. The vaults of King's College Chapel, Henry VII Chapel at Westminster, Bath and Peterborough, to name the four vaults which span 8 m or more, all testify to the skill and basic understanding of their architects.

It is, of course, not necessary to pursue the mathematics of the ideal profiles in order to understand the mechanics of the fan vault, or to obtain estimates of significant quantities, for example the horizontal thrust. The slicing technique, coupled with Hooke's hanging chain, gives an immediate entry into the problem. Figure 4.37 is based on a cut in Mackenzie's (1840) account of the construction of the roof of King's College Chapel. Mackenzie imagines stones to be assembled to form an inverted concave conoid, and fig. 4.37 shows the structure partially completed (Mackenzie notes that the base must be prevented from spreading). On the principle of the dome, he remarks, the last completed horizontal circle of masonry will stand secure.

The added curved heavy lines in fig. 4.37 represent flexible chains, each one loaded with the masonry contained within the corresponding 'orange' slice. Each chain is of course in tension, and acts independently of its neighbours, so that a satisfactory system of statics could be constructed for a half (or indeed a quarter) conoid. Moreover, three powerful conclusions emerge from a study of fig. 4.37.

At their lower ends the cables, in tension, must pull against some abutment, and this is shown schematically in fig. 4.37 as a curved compression ring. In the real structure, corresponding to the inversion of

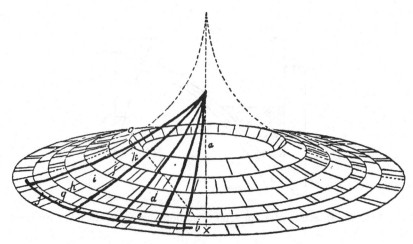

Fig. 4.37. An inverted concave conoid (from Mackenzie 1840; heavy lines added).

fig. 4.37, the 'chains' will be in compression, and the edge of the fan must be 'prestressed' in compression, as has been seen – by the fans leaning against each other at midspan, and by being required to support the heavy spandrel masonry (including half-tonne pendant bosses at King's College Chapel).

Second, fig. 4.37 indicates that the 'flexible lines' lie within the masonry at the wider portions of the fan but that, as the central axis is approached, the geometry of the lines may be expected to differ markedly from that of the fan itself. The solid fill of the vaulting conoids is essential for stability.

Finally, the flexible lines meet on the axis of the conoid and will impose a resultant concentrated pull at a certain point on that axis. In the real structure the fan vault will deliver a more or less concentrated thrust at the side walls of the Chapel; the calculations may be modified to take into account the weight of fill in the conoid. If these calculations are carried through for King's College Chapel, it may be deduced that the horizontal thrust at each buttress is about 10 tonnes, rather than the 16 tonnes quoted above for the 'ideal' profile.

5

Some Structural Elements

5.1 The wall

The three basic assumptions about the behaviour of masonry – no tensile strength, infinite compressive strength, no slip of the stones – have effectively eliminated material properties from the discussion. The previous chapters have attempted the analysis of arches, domes and vaults from, so to speak, a somewhat distant viewpoint; it is the overall shape of these structures, rather than their detailed construction, which controls structural action. Such geometrical considerations are also paramount for the study of the flying buttress made at the end of this present chapter, but a closer view must be taken of, for example, the wall, if its action (and its pathology) are to be understood.

Some elementary geometry is of course involved even in the idealization of a wall as a flat-sided slab of uniform thickness. It seems obvious, for example, that the wall must not be too thin compared with its height and length, and this intuition can be supported easily by rational argument. A thin wall may be built vertically and remain upright under the action of its own weight, but it must have some reasonable margin of safety against settlement and accidental tilt of its foundations. The centre of gravity of a cross-section of the wall must not move outside the verticals drawn through the limits of its base, and a certain minimum ratio of thickness to height will give the required margin. Equally, a free-standing wall of sufficient thickness must not be blown flat in a high wind, and this again leads to a (calculable) minimum thickness of construction.

A free-standing wall of sufficient thickness needs no other support. A thinner wall of great length may, however, be stabilized by buttressing; effectively, the wall is thickened at various points along its length. If the wall forms part of a building, say the wall of an aisle of the church, then the buttresses will serve also to absorb lateral forces arising from the timber roofs or masonry vaults.

The assumption of infinite compressive strength must, however, be questioned in the analysis of the behaviour of a thick wall. It is true that mean stresses in walls are small, so that there should be no danger of crushing of the material, but the calculation of a mean stress can be misleading. The medieval construction of a wall, perhaps 1 m thick, or even 2 m for some great towers, often consists of two outer skins of good coursed ashlar (the skins being say 200 to 300 mm thick), with random rubble and mortar fill contained between these skins. It is common, for reasons noted below, for the rubble fill to slump, so that the inner and outer skins are forced to carry the whole of the load on the wall, with a consequent great increase of stress in these load-bearing skins. Some aspects of the behaviour of the masonry under these conditions are noted in the next section of this chapter, where the behaviour of piers is discussed.

The fact is that a construction of skins and rubble fill has little inherent tensile strength across the thickness of the wall, and two devices were commonly used during building to try to provide reinforcement. The first, which provides little connexion between the skins, consists of longish baulks of timber laid horizontally in the central core of rubble and mortar. The second involves the use of 'through-stones', larger blocks of masonry passing right through the wall, and connecting the skins at more or less regular intervals in the construction. For walls up to 1 m thick this is a practical and effective way of providing cohesion.

For thicker walls, sufficiently large pieces are not in general available to act as through stones. There will then be no positive connexion between the two faces of the wall, and the skins will tend to drift apart. This secular tendency to drift, which affects all parts of a masonry structure, has a particularly simple explanation with reference to the kind of wall being considered. An initially tight construction will be subject to thermal strain, and to vibration from ox-carts, container trucks and sonic booms, all of which will cause fissures to open within the fabric, or to extend cracks which may have developed already from the shrinkage of drying mortar. These same vibrations, afforced perhaps by the playing of the organ, by bell-ringing, and by buffeting from the wind, will cause dust to settle in the open cracks and prevent their closure. A renewed cycle of temperature change, and other attacks from an essentially hostile environment,

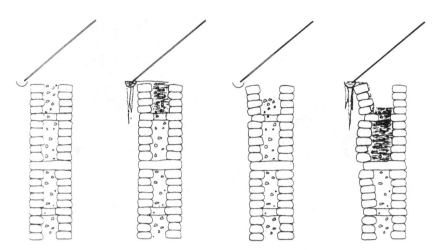

Fig. 5.1. The drift apart of masonry skins (from Beckmann 1985).

will then cause these developed cracks, and other new cracks, to open further, and these in turn will be filled with debris to prevent their closure.

A ratchet mechanism is in operation; the overall dimensions of a masonry structure can only grow, never decrease. Eventually the geometry may alter to the extent that the changes are no longer 'containable within the thickness of a pencil line on the drawing board', and remedial action must be taken. Occasionally the ratchet mechanism may be driven positively, as noted by Poul Beckmann (1985); fig. 5.1. A leaking gutter can allow water to penetrate the rubble and mortar fill; if this water then freezes, it will expand, and the two skins will be pushed (rather than drift) apart. As before, a thaw will not pull the skins together again; rather, the fill will slump, and the defect becomes progressive.

Treatment of the defect is not easy. Voids in the fill can be injected with grout; this will close existing cracks, but does not provide a tensile connexion through the thickness of the wall, and renewed drift will occur. The essential requirement for stabilization is to connect together in a positive way the two skins of the construction, and metal reinforcing rods can be used, grouted into holes drilled through one skin and ending in the other. The defect assumes particular importance when it arises in the four walls of a tower, and the question is referred to again in the next chapter.

Fig. 5.2. Fractured wall of the church of Sainte-Geneviève, Paris (the Panthéon) (from Rondelet 1834).

5.2 The pier

Attached piers may be regarded as local thickening of walls, required mainly for the geometrical stability of the construction (i.e. if they are not in some sense purely decorative). The case is very different for arcade piers, and above all for crossing piers which carry the weight of a tower.

The typical large medieval crossing pier is constructed, like the thick wall, from a well-cut stone skin and a rubble and mortar core. However, unlike the wall, the skin of the pier may not tend to drift out; on the other hand, shrinkage of the mortar will, as with the wall, lead to voids in the core, and to a possible slump of the internal material. Since a crossing pier already carries a high mean stress, any loss of effective cross-section is a serious matter; the load on the pier is thrown on to the skin and very high stresses can be engendered.

Two consequent local mechanisms of distress can be observed. In the first, pressure points force off wedges of stone, and roughly triangular spalled areas may be seen in the skin. Figure 5.2 shows a sketch of a wall which has been overloaded and exhibits defects of this sort. Alternatively, high pressures between the stones can lead to more or less *vertical* splitting of the stones in the skin. This type of fracture can also be seen in fig. 5.2.

The illustration is due to Rondelet (1834), who was in charge of the construction of the church of Sainte-Geneviève (now the French Panthéon) from 1770 onwards. Defects similar to those illustrated developed when load first began to be imposed on the four crossing piers. The full final load on a single pier is over 34 000 kN (i.e. the total weight at the crossing approaches 14 000 tonnes), and each pier has area of about 12.6 m^2, so that the *mean* stress is 2.7 N/mm^2. However, the construction of the piers might have been designed purposefully to lead to much higher stresses; the stones were carefully dressed at their bearing surfaces for about 100 mm, and then cut away behind this border. In this way very tight (3 mm) joints were achieved without the need to dress carefully the whole surface of the stone, and mortar was stuffed through the joints in an attempt to fill the internal cavities. Even if the attempt was in fact successful, the mortar shrank on drying, and the whole load on each pier was carried not on the full area of 12.6 m^2, but on an effective peripheral area of 2.3 m^2.

Thus a calculated mean stress of 2.7 N/mm^2, about 10 per cent (actually slightly greater) of the crushing strength of the stone, was replaced in practice by an actual stress of say 15 N/mm^2, now close to the failure point, and defects similar to those sketched in fig. 5.2 were inevitable. Although the defects are unsightly, the construction was not in danger of collapse; the spalling of the surface would have brought more of the cross-section of the pier into action, and stresses would have been lowered. Indeed, an astonishing remedy was put in hand in 1779, and this had some success; masons spent more than a year opening all the horizontal joints with a saw in an attempt to throw the load away from the surfaces of the piers towards their centres.

This example from eighteenth-century France is one of bad practice, and was condemned as such at the time. The medieval construction of skin and rubble fill is, with hindsight, also bad practice, but it was universal. As a result stresses in the skins of crossing piers would always have been high, and 'drift' of the whole fabric would only increase those stresses. It is to be expected, then, that any defect betraying a high stress, say a vertical split in the masonry, will tend to increase slowly, and there will come a time, at perhaps the 500 years from the date of construction mentioned in Chapter 2, when some major intervention may be needed. (As noted in Chapter 2, intervention came too late at Chichester in 1850, if indeed it was failure of the masonry by stress rather than geometrical distortion due to settlement that led to collapse of the tower.)

The strengthening of a crossing pier, carrying perhaps 3000 tonnes, is a delicate operation. Complete removal and replacement of the piers seems difficult; falsework to carry the load temporarily can be designed, but that falsework must transmit its load to the ground, either through the existing foundation to the pier, or through newly provided footings. If indeed these difficulties can be overcome, then the falsework could be jacked until all the load is transferred out of the pier; this could then be rebuilt in solid masonry at leisure. There seem to be no large-scale examples of such a drastic remedy.

At Milan, however, in the early 1980s (i.e. nearly 600 years after their construction), the four crossing piers were entirely rebuilt using a different technique. Each pier was encased in longitudinal steels and girdles, and a wedge (a 'cheese slice') was removed by cutting, the wedge extending to the centre of the pier. New stone was inserted to replace the removed material, and the whole cross-section was gradually renewed throughout its length by cutting successive slices.

A technique requiring a minimum of falsework, but still a delicate operation, is that of grouting. As with the consolidation of a wall, it is of no use to introduce grout without at the same time ensuring that the final fabric of the pier has some tensile strength, to guard against future drift. Thus reinforcing rods (stainless steel) may be inserted, in a logical pattern, in a large number of holes drilled virtually right through the pier. The crossing piers at Worcester have been strengthened in this way (1990–91), and traces of the work are almost invisible even at the closest inspection of the surfaces.

5.3 The pinnacle

Both the wall and the pier must have appropriate dimensions in order that they may carry structural forces, but the discussion in the last few pages has been concerned more with internal material properties rather than overall structural action. One of the celebrated disputes of architectural history of the first years of the twentieth century concerned the function and mode of action of the pinnacles placed on the buttresses of French Gothic (as for example those shown in Viollet-le-Duc's drawing of Amiens, fig. 5.16); are pinnacles structurally necessary, or are they in some sense purely decorative? The dispute sprang from the reaction against the *rationalisme* imputed by Viollet-le-Duc to the whole of the Gothic structure – every element of the structure must have a purpose, whether it be to hold the building together or to allow rainwater to escape.

Abraham attacked this functional view in his monograph of 1934; he demonstrates some correct (and some incorrect) structural analysis, but the whole of his long essay is distorted by his desire to prove his paradoxes that, of the two *essences* of Gothic, the rib vault and the flying buttresses, the rib does not carry the vault and the flying buttress does not thrust. (The rib was discussed in Chapter 4 – its function did indeed give rise to another celebrated dispute. That the flying buttress does in fact thrust will be seen later in this chapter.)

Paul Frankl (1960), in his massive study of Gothic, discusses the *rationalisme* of Viollet-le-Duc.

Every form has its practical purpose. The pinnacles are an important example of this, 'which by their weight give the buttresses (*contre-forts*) all the stability (*fixité*) necessary to support the thrust of the flying buttresses'. From this thesis, traceable to Sir Christopher Wren, one would have to conclude that the flying buttress would give way and consequently the nave vaults collapse if the pinnacles were to be removed. Paradoxically, according to this thesis, the more the piers are weighted down vertically by a superstructure, the thinner they can be made.

Frankl is here confusing stability with strength. He seems to imply that the weight of a pinnacle will cause distress to a pier, perhaps causing crushing of the masonry; but Viollet-le-Duc's word '*fixité*' is clear, and has been properly translated by Frankl as 'stability'. There has been no mention of the stresses in the masonry; the stresses in the external buttresses are low, and even a massive pinnacle will increase their small values by only a further small amount. In practical terms, the presence of a pinnacle will have negligible effect on the overall strength of the supporting masonry, but it may have a slight beneficial effect on the stability of that masonry.

The general stabilizing effect of a pinnacle was well understood early in the nineteenth century; for example, Moseley's book of 1843 discusses lines of thrust in masonry, 'the stability of a pier of buttress surmounted by a pinnacle', and, specifically, 'the Gothic buttress'. Moseley refers continually to earlier French work, and Viollet-le-Duc must have been well aware of the work of these French engineers at the *École Polytechnique* and the *Ponts et Chaussées*. Abraham's sneer cannot be accepted ' ... that neither flying buttresses nor pinnacles were necessary. Many a French cathedral had none, and acquired them only when restored by Viollet-le-Duc'.

Yet there is some excuse for the doubts about the efficacy of pinnacles. It seems to the eye, and this is confirmed by calculation, that a pinnacle can have little effect on the overall stability of a pier. Abraham is right when he says that it is a fine pinnacle that weighs a hundredth of the total weight on a pier. The main effect of a pinnacle is indeed not concerned with overall stability, but is localized at the head of the pier.

Figure 5.3 is a schematic diagram based upon the buttressing system at Lichfield. In fig. 5.3(a) the flying buttress delivers an inclined thrust

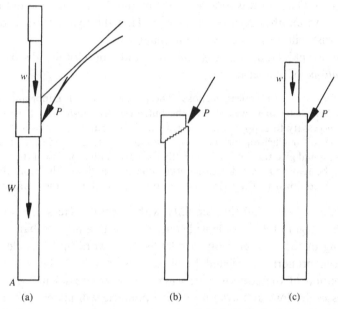

Fig. 5.3. Forces acting on a buttressing pier (Lichfield Cathedral, schematic). (a) The buttress will be stable if it does not overturn about toe *A*. (b) Sliding failure in the absence of a pinnacle. (c) Sliding failure prevented.

P to the external pier, and this is surmounted by particularly massive masonry. In fig. 5.3(b) this superimposed masonry has been removed, and it may be imagined that there will be a tendency to sliding failure under the action of the thrust *P*. To avoid such failure, a simple expedient is to add a relatively small weight *w* (fig. 5.3(c)), in order to increase the frictional force along a potential line of slip. The calculations are simple (and were given for example by Moseley (1843)). It turns out that the line of action of the added weight *w* is immaterial. Thus there is no objection, from this point of view, to placing the pinnacle towards the outside of the main buttress, where its small effect on overall stability would be even further diminished.

The pinnacle is an example, then, of design to satisfy an apparently subsidiary structural criterion. As has been seen, the stability of masonry is assured in the main by a correct overall geometry of the structure, but this assurance depends on the fact that the three basic assumptions about material behaviour are satisfied. The assumption of zero tensile strength is clearly conservative. The assumption of infinite compressive strength, that is, of a generally low level of stress, may have to be examined critically, as for example for a wall or pier with a weak core.

The assumption about slip not occurring in the masonry is one that is usually satisfied in practice. In particular circumstances, however, a low level of general compressive stress may be necessary in order to mobilize sufficient friction to prevent slip. The pinnacle is an example of an effective compressive prestressing agent.

5.4 The flying buttress

The north and south walls of a church must be buttressed against the thrusts imposed at a high level by the masonry vault. (As has been noted, thrusts in the east–west direction are transmitted from bay to bay of the vault, and must be resisted finally by the masonry at the east end and by the westwork.) If there are no side aisles, then the main buttressing piers may be placed directly against the walls (King's College Chapel, fig. 5.4; the Sainte-Chapelle, fig. 5.5). If, however, there are side aisles, then the thrusts must somehow be conveyed above the aisle roofs; the flying buttress is, effectively, a compressive prop passing between the vault and the external main buttress.

It was this visual exposure of the masonry structure that offended the 'Beaux-Arts' critics of Gothic at the turn of the century – Guadet, for example, called Gothic the propped-up style of architecture. Indeed, the flying buttresses of Notre-Dame in Paris, spanning *two* side aisles, have enormous visual impact. They are the price that is paid, if in fact one agrees that a price is paid at all, for the Gothic dissolution of the wall. The 'engineering' structure can no longer be hidden within the elements of pier and vault, but escapes from the internal architecture of the building.

The flying buttress was used as a prop against thrusts other than those generated by the high vault. Figure 5.6 (from Jackson 1906) shows a cross-section of Westminster Abbey, and it will be seen that, on the north (right-hand) side, there are two flying buttresses. The main external pier is placed outside the north aisle, and the lower of the two flyers is positioned well to receive the thrust from the high vault, and transmit it across the triforium. The upper flying buttress does not help in sustaining the vault, and its function has been discussed by Fitchen (1955).

There are in fact two main functions of the upper flyer. The tall timber roof is subject to lateral wind forces; at Westminster, the horizontal thrust

Fig. 5.4. Cross-section of King's College Chapel (1448–1515) (From Mackenzie 1840).

Fig. 5.5. Cross-section of the Sainte-Chapelle, Paris (1244–47) (From Viollet-le-Duc 1858–68).

at the parapet due to wind may reach about 7 tonnes in each bay. Thus the upper flyer, placed at parapet level, will receive this thrust and convey it (as does the lower flyer the vault thrust) to the external pier and so to the ground. Further, the timber roof itself is liable to distortion, and, even with a tie such as that shown in fig. 5.6, it may spread and impose a lateral load on the side walls at parapet level. Without a tie (and the construction of the vault often precludes the use of a tie), the rafters will exert direct lateral pressure on the walls.

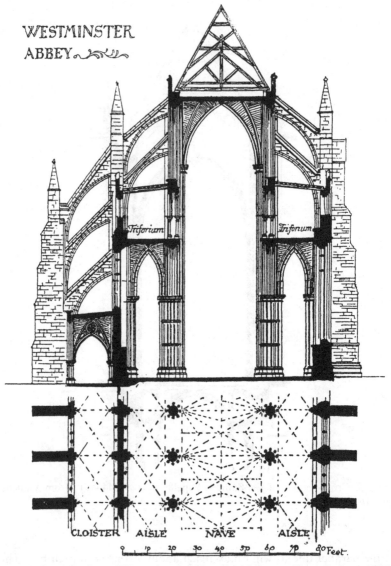

Fig. 5.6. Cross-section of Westminster Abbey, looking west (from Jackson 1906).

Such a double propping system was usual for the great cathedrals of French Gothic, and Westminster Abbey is of course a French church. Figure 5.6 shows *three* props on the south side of Westminster Abbey; this side of the church is abutted by a cloister, which prevents the main external buttressing pier being placed directly against the aisle wall, as on the north side. Thus the two upper props on the south side, at parapet and high vault level, transfer their forces through an intermediate pier to the external buttress. The thrust from the south aisle vault is not, however, resisted directly by this buttress, as it is to the north, and this dictated the use of the third flying buttress over the cloister.

All these flying props, then, convey thrusts from the main fabric of the church to the external buttressing piers. A thrust line may be imagined to pass through space above the aisle (or above the cloister in fig. 5.6), and this thrust must be surrounded with masonry. If the flying buttress were weightless, then its best form, by analogy with the weightless cord in tension, would be a straight line. Gaudí adopted this logical solution in some of his construction (for example the Colonia Güell chapel). The typical Gothic flying buttress has indeed a flat extrados, and it has been noted, fig. 2.4, that this flat-arch form has no mechanism of collapse.

In order to understand in detail the way a flying buttress acts, the techniques outlined in Chapter 2 may be used. The analysis of masonry arches was well understood in the nineteenth century, and engineers were able to proceed to orderly design of their structures. The work, starting in France in the eighteenth century, was mainly done in Germany and Switzerland; Ungewitter (1901), whose discussion of vaults has already been noted, paid considerable attention to the flying buttress in his two-volume study of the construction and behaviour of Gothic, using the now established techniques of analysis. His plate 41 comprising figures 402–410 is reproduced in fig. 5.7. (Figure 408 shows a complete pier of a church, acted on at the top by the weight of the parapet and the thrust of the high vault and resisted by the (single) flier, and lower down by the thrust from the aisle vault.)

Figure 402 shows two possible lines of thrust lying wholly within the masonry; for the lower thrust line Ungewitter has divided the flying buttress into seven segments and, with the aid of 402 a, he has constructed the funicular polygon, that is, the shape of the (inverted) chain (i.e. the thrust line) which will equilibrate the weight of the flying buttress. The upper line is constructed similarly (details not shown by Ungewitter) and corresponds to a higher value of horizontal thrust being imposed on the buttress. Just as for the masonry arch of Chapter 2, there is a whole

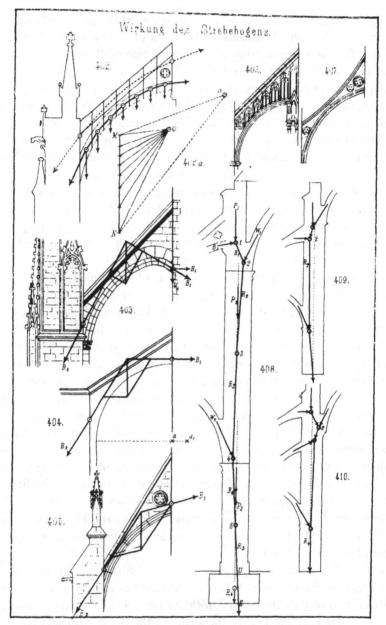

Fig. 5.7. The action of flying buttresses (from Ungewitter 1901).

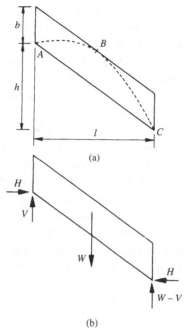

Fig. 5.8. The inclined flat arch (as a flying buttress).

range of thrusts for which the flyer will be stable; how the variations can arise will be seen when the detailed analyses of particular flying buttresses are made.

Figures 403, 404 and 405 show, in general terms, that the direction of the thrust (B_1) imposed on the wall (of the nave or choir) depends on the shape given by the builder to the flying buttress. The late (Flamboyant) buttress of Figure 403 (St Ouen, Rouen) was constructed without that firm intuitive grasp of structural principles shown in the twelfth and thirteenth centuries. It is of course the thrust B_1 that is required to counteract the outward and downward thrust of the high vault, and the excessive masonry at the head of the flyer in Figure 403, leading to the downward droop of the thrust line, indicates an uneconomical (although finally effective) design.

The broad picture of the behaviour of flying buttresses may be studied with reference to the inclined flat arch of fig. 5.8; this is a crude model of the real structural element, but serves nevertheless to indicate some features of the structural action. The analysis may be approached in the same way as that for the voussoir arch – the centering supporting the

flying buttress is removed, the buttress thrusts against the nave wall and
the external pier, and one or other of these (in fact the pier) gives way
slightly so that the span l increases. Three hinges will form at A, B and C
in fig. 5.8(a); the broken line shows the approximate position of the line
of thrust; cf. fig. 4.21(a). At sections A and C the line of thrust passes
out of the flying buttress into the nave vault and the external buttressing
pier, respectively.

An immediate problem is the determination of the location of the
hinge at B where the line of thrust touches the extrados. There are
various methods of attack for this problem, which is essentially one of
geometry, including the use of graphic statics as illustrated by Ungewitter
in fig. 5.7. For the flat arch of fig. 5.8 analytical methods may be used,
and it turns out that in this idealized case the point B lies at midspan
of the flyer. Using this fact, the values of the forces V and H marked in
fig. 5.8(b) may be determined; these are the forces necessary to keep the
flyer of weight W in equilibrium. Using the symbols shown in the figure,

$$H = \frac{Wl}{8b} \tag{5.1}$$

$$\text{and} \quad V = \frac{1}{2}W\left(1 - \frac{1}{4}\frac{h}{b}\right). \tag{5.2}$$

While these results have been obtained for the parallel-sided flat arch,
certain general features may be noted. First, the values of V and H
displayed above represent the *passive state* of the buttress; the buttress
leans against the nave wall with this minimum value of $H = Wl/8b$, and
stability cannot be maintained with a horizontal thrust smaller than this.
Any slight inward shift of the nave wall, or outward shift of the main
buttress (i.e. a further increase in span of the flying buttress), will take
place with V and H remaining constant.

Second, the value of H depends only on the span l and depth b of the
buttress, and not on its inclination (measured by h). Third, for a steeply
inclined buttress ($h > 4b$ in this model), equation (5.2) shows that the
value of V becomes negative; the sort of drooping thrust at the head of
the buttress illustrated in Ungewitter's Figure 403 (fig. 5.7) corresponds
to a positive value of V.

Finally, numerical values will give some idea of the forces involved.
A 45° flyer might have $h = l = 6$ m; if b is taken as 1.5 m, then an
appropriate width of the flyer would give a total weight of material of
say $W = 10$ tonnes. Thus H is determined as 5 tonnes from equation

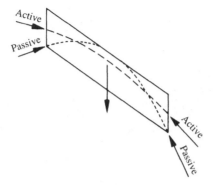

Fig. 5.9. Passive and active states of an idealised flying buttress (cf. fig.5.8).

(5.1) and V as zero from equation (5.2); the thrust line in fig. 5.8(a) is just horizontal at A where the flyer meets the nave.

A weight of flying buttress of 10 tonnes might be typical of that of a large cathedral (e.g. Clermont-Ferrand, below), but the passive thrust, that is, the minimum value of H in fig. 5.8 necessary to ensure stability, will be rather less than 5 tonnes. This is because, in the real shape of a flying buttress, some unnecessary material is cut away from the parallel-sided form pictured in fig. 5.8. That this is possible is at once evident from the position of the passive thrust line in fig. 5.8(a). The material in midspan, below the point B, is not being used to carry the thrust and is, in this passive minimal state, effectively so much dead weight.

However, a flying buttress does not of course work in its passive minimal state. It is called upon to resist the steady permanent outward thrust from the high vault, and, as was illustrated by the typical numerical example of chapter 4, this thrust might have a value of 15 or 20 tonnes. Thus the active state of the same schematic parallel-sided flyer might be as shown in fig. 5.9; an active line of thrust is shown which equilibrates the vault thrust and lies wholly within the masonry. There is no theoretical way of determining the 'actual' position of this thrust line, although inspection of a flying buttress may reveal cracking and so locate at least some sections through which forces may be transmitted. However, as for the voussoir arch of Chapter 2, it is necessary only to show that the thrust line lies within the masonry to be assured of the stability of the construction.

Figure 5.10 shows Viollet-le-Duc's drawing of Clermont-Ferrand, which has the two-tier system of buttressing. Note that the buttresses receive vertical support from the piers placed between the main external buttress

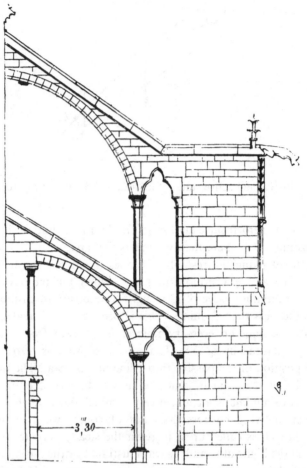

Fig. 5.10. The buttresses at Clermont-Ferrand (from Viollet-le-Duc 1858–68).

and the nave wall; the vault thrust (from the lower flyer) and the wind load (from the upper flyer) will act effectively low down the external buttressing pier, and only the most rudimentary pinnacle is seen in the figure (cf. fig. 5.12 for Notre-Dame, Paris). The upper flying buttress will be analysed using the technique of the funicular polygon.

The work is simple, and has already been described in the discussion in Chapter 3 of Poleni's analysis of the dome of St Peter's. A sketch is made of the buttress (fig. 5.11), and divided into a number of sections whose weights are calculated. The shape of the 'hanging chain' is then found by constructing a force polygon (cf. Ungewitter's Figure 402

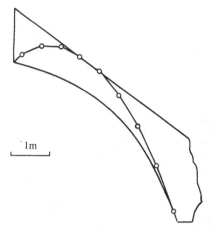

Fig. 5.11. Passive line of thrust for the upper flyer, Clermont-Ferrand.

in fig. 5.7), and it remains to locate this line of thrust on the outline drawing. The passive line of thrust will pass through the intrados at either end of the flyer (cf. points *A* and *C* in fig. 5.8(a)); the hinge point at the extrados is found by trial and error, and the correct position of the passive line of thrust is as shown in fig. 5.11. Once the line of thrust has been located, the forces necessary for equilibrium may be calculated, using the known weights of the sections of the flyer. For a total weight of flying buttress of about 10 tonnes, the horizontal thrust is found to be about 3 tonnes; the upper flying buttress at Clermont-Ferrand leans against the parapet masonry with a minimum force of this value.

Fitchen (1955) estimates that the wind acting on the tall timber roof of Clermont-Ferrand could produce a force of 20 tonnes per bay of the nave. Thus the upper flyer in its active state will have a line of thrust in some such position as that shown schematically in fig. 5.9.

Figures 5.12 and 5.13 give a similar analysis for the great flying buttresses of Notre-Dame, Paris. These buttresses span two side aisles, and are in fact a thirteenth-century replacement for the original twelfth-century construction. The passive thrust in Notre-Dame is four to five times as great as that of Clermont-Ferrand, since the span is so much greater, but the thrust lines in figs. 5.11 and 5.13 are very similar in shape.

As a final example of conventional flying buttresses those at Lichfield may be examined. Figure 5.14 shows a cross-section of the cathedral

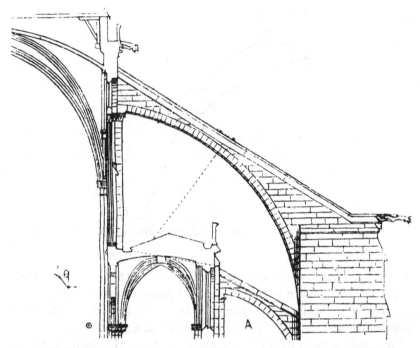

Fig. 5.12. The thirteenth-century flying buttress at Notre-Dame, Paris (from Viollet-le-Duc 1858–68).

(cf. fig. 5.3 in the discussion of the role of the pinnacle), and fig. 5.15 shows the computed passive line of thrust. The associated numerical values give insight into the reserves inherent in the standard flying buttress supporting a high vault. The passive thrust at Lichfield is about 3 tonnes; the maximum value of active thrust sufficient to cause crushing at the smallest cross-section of the flier is about 1000 tonnes. (Since a straight line can be passed from one end to the other of all the buttresses sketched in figs. 5.11, 5.13, and 5.15, there is no possibility of mechanisms of collapse being formed.) If a factor of 1/10 is taken as the value of thrust to cause crushing, it may be concluded that the flying buttress at Lichfield can work comfortably at thrusts anywhere between 3 and 100 tonnes; the flyers will adjust themselves, automatically and exactly, to resist any value of thrust, live or dead, within that range.

Thus a flying buttress at Lichfield will actually be subjected to a steady thrust greater than the minimum value of the passive state, and will resist exactly the steady value of the vault thrust. In addition, the

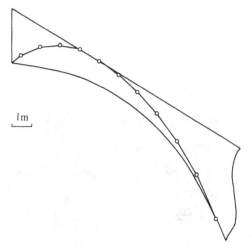

1 m

Fig. 5.13. Passive line of thrust for Notre-Dame, Paris.

flying buttresses at Lichfield will resist live wind loads. As has been remarked, the passive state with its implied three hinges in the flyer is not likely to be observed in practice, since the vault thrust will usually be substantially greater than the minimum passive thrust of the flyer. Under exceptional circumstances, however, when a flyer is stronger and heavier than necessary and the external main buttresses may have suffered settlement or tilt, then the hinge pattern may be seen, as recently at the Chapter House, Lincoln.

There are many variations of the 'standard' forms of figs. 5.10, 5.12 and 5.14, from the almost horizontal fliers of Palma di Mallorca to the steep props at Bath and Cirencester. All, however, have the property that straight lines of thrust which lie within the masonry may be drawn from one end to the other, so that the strength of the fliers is limited effectively only by the crushing strength of the material.

Occasionally this requirement was ignored, as at Amiens, or perhaps not perceived. Figure 5.16 shows the buttresses at the chevet, as they were built and as they still exist. Similar buttresses were provided originally to the nave, but these failed by upward buckling, and were replaced by a different design in the fifteenth century. There are really two buttresses in fig. 5.16 – the curved lower rib, positioned to transmit the steady vault thrust, and the straight upper rib, normally unloaded, but with an essential function as a brace against wind forces. The two ribs are

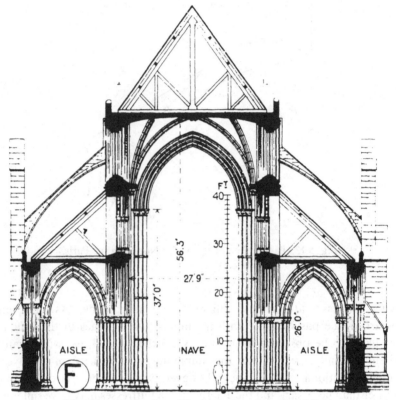

Fig. 5.14. Sir Banister Fletcher's drawing of Lichfield.

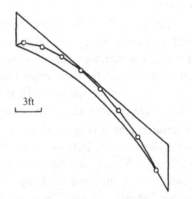

Fig. 5.15. Passive line of thrust for Lichfield.

Fig. 5.16. The buttressing system at Amiens (from Viollet-le-Duc 1858–68).

connected by tracery, so that the weight of the upper rib is carried by the lower.

The curved lower rib is an arch requiring a minimum horizontal thrust of 5 tonnes for stability; the usual numerical analysis can be made, taking into account the weight of the upper rib, and fig. 5.17 shows the passive line of thrust. The other line of thrust in the figure corresponds to the largest active horizontal thrust that can be sustained if the rib is to remain stable. If a straight line could be drawn from end to end of the rib, the thrust could reach a very high value; as it is, the curvature of the lower rib at Amiens limits the value of thrust to almost exactly four

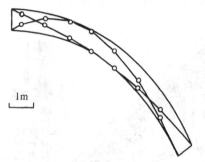

Fig. 5.17. Passive and active lines of thrust for the lower rib at Amiens.

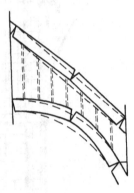

Fig. 5.18. Possible mode of failure, Amiens.

times the passive value. Thus the maximum value of horizontal thrust that can be sustained is 20 tonnes, and this is about equal to the vault thrust at Amiens.

The curved rib, forced by its geometry to operate within the limits of 5 and 20 tonnes, buckles upwards under the action of the vault thrust. The tracery connexion ensures that the straight rib is also pushed aside (fig. 5.18), allowing the nave wall to lean outwards. This was possibly the mode of failure at Amiens, which perhaps proceeded progressively under the added action of wind.

5.5 The tas de charge

The equilibrium of the whole Gothic structure is ensured by the presence of the flying buttress, which provides the counterthrust to the steady thrust of the high vault. It was seen in Chapter 4 that the visual

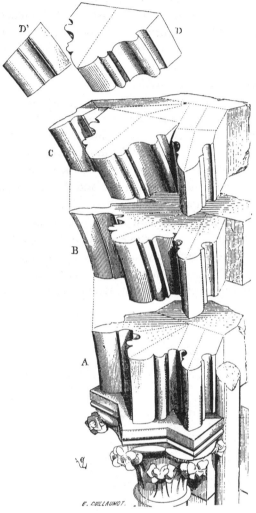

Fig. 5.19. Exploded view of masonry courses adjacent to springing of a vault (from Viollet-le-Duc 1858–68).

geometry of the ribs of the vault is structurally misleading – the thrust does not follow the ribs, but 'escapes' from the masonry of the vault as the springing is approached, and passes into the vaulting conoid. Thus, from the inside of the church, the vault appears to spring from a certain level, but the flying buttress is positioned at a higher level on the outside wall of the church (see, for example, figs. 4.18 or 4.20).

It was noted also that the vaulting conoid must be provided with fill to provide a passage for the vault thrust. Gothic builders did not rely entirely on the rubble and mortar in the conoid for this passage; instead, it is usual to find a through-stone, the tas de charge, connecting the head of the flying buttress on the outside of the church with the solid masonry of the ribs and webs on the inside.

In addition to these structural functions of the tas de charge, that is, of binding the work together and of ensuring a firm passage for the vault thrust from the inside to the outside of the church, there is also a constructional function, which may be appreciated from Viollet-le-Duc's drawing of the masonry courses adjacent to the springing (fig. 5.19). For this quadripartite vault the diagonal ribs and the main transverse rib run together above the capital of the pier, and the mouldings of all three ribs can be cut from a single stone. Single blocks can also be used for higher courses, and the topmost such stone, running through the fabric to the head of the flying buttress, may form the tas de charge. All these lower stones may be positioned without formwork; when separate stones are cut for each rib, however, then constructional formwork is necessary for support.

6

Towers and Bells

The title of this chapter is that of a Handbook (Frost 1990) compiled by the Towers and Belfries Committee of the Central Council of Church Bell Ringers. The Handbook concentrates of course on matters related to bells, but it is also an invaluable source of information regarding the structure and maintenance of masonry towers. Naturally enough, it does not discuss the kind of large-scale disaster referred to in Chapter 2 – the overall collapse of towers such as those at Winchester, Gloucester or Worcester, or the hasty insertion of props at Wells. These disasters occurred within the soil-mechanics time scale for consolidation, of up to twenty years (although, as has been noted, some towers have survived for much longer periods before mysteriously collapsing). It seems evident that settlement of foundations leading to tilt may be a cause of potential danger for a tower; many do, in fact, survive the original high-risk period in which settlement occurs, and appear to survive at alarming angles of inclination (the eye is sensitive to very small deviations from the vertical). Perhaps the most famous example is the campanile of Pisa, but there are many others in Italy, particularly in Venice and in the islands of the lagoon.

6.1 The Campanile in Venice

The collapse of the Campanile in Venice, on 14 July 1902, received extensive contemporary discussion and analysis (see e.g. Beltrami 1902), particularly since the final phase, from the instant that it was known that the 13 000 tonnes of masonry would collapse, to the actual event, lasted 3 days and 19 hours. The disaster was indeed closely observed, and it was not apparently accompanied by any tilt of the tower; rather, fissures were seen to widen, and the final pile of rubble offered few clues as to the cause of the defects. Similarly, there are no reports of tilt at Ely in 1322, when the crossing tower collapsed two centuries after it had been built (and was at once replaced by the octagonal timber vault and lantern), or at Chichester in 1861, after seven centuries of seemingly comfortable existence.

Certainly examination of the foundations of the Campanile after the fall of 1902 revealed that geotechnical failure leading to tilt was an unlikely cause of collapse. It would seem that some other defect can be engendered by uneven settlement, and some clues are afforded by an examination of the history of the Campanile and of its condition before the fall. For example, vertical fissures were known to have been present in the fabric of the tower. Equally, the Campanile, in its long history of over a thousand years, had experienced previous distress and partial destruction. Thunderstorms played a part; the structure was struck by lightning in 1388, and again in 1417 and 1489, and on this last occasion it was virtually ruined. Lightning again damaged the tower severely in 1548, 1565 and 1653; in 1745 it was again almost destroyed, and 37 fissures had to be repaired. There was yet further damage in 1761 and 1762; in 1766, however, a Franklin lightning rod was installed, and the tower was relatively unharmed until 1902. (A similar Franklin rod was installed in Wren's St Paul's in 1769.)

The four walls of a medieval tower are constructed as described in Chapter 5 – two skins of ashlar may conceal a rough fill of rubble and mortar. Fissures can develop in this central fill as the two skins tend to drift apart, although in fact the four internal skins are constrained by each other to remain more or less in place. However, there is nothing to prevent the outer skins from moving, other than the intrinsic weak tensile strength of the material. (Correction of this tendency is discussed later in this chapter.) The type of cracking induced by this drift lies parallel to the faces of the walls.

The vertical fissures observed in the Venice Campanile were at right angles; they occurred in a plane through the thickness of the wall. A material with no tensile strength has no shear strength, and such vertical cracking can be promoted by slightly uneven foundation settlement taking place during consolidation of the soil. Figure 6.1 gives a sketch of the plan of a corner of a tower, and it will be seen that there is a possibility of the corner becoming detached. Robert Willis's account makes it clear that something of the sort preceded the collapse of the crossing tower at Chichester in 1861. A vertical crack had existed at the corner of the Campanile for at least a century, and it was the widening of this crack

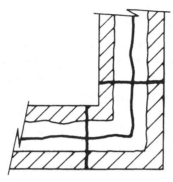

Fig. 6.1. Vertical cracking in the walls of a tower (schematic). The drift apart of the 'skins' can lead to cracks in the rubble fill; settlement can lead to cracks through the thickness of the walls, leading to isolation of one corner.

over nearly four days that led to the final downfall. In fact, this crack connected a series of eight windows placed in the corner of the tower to light a stair, so that there was an easy path along which a fissure could develop.

Vertical fissures will become wet in thunderstorms, and will provide good conducting channels for the 30 000 amperes associated with a lightning stroke (Schonland 1964). The corresponding temperature rise can be 15 000°C, and pressures generated by the virtually instantaneous production of steam can damage severely the overall fabric of a tower. (In the same way the sap-wood channel, acting as a conductor, can lead to the explosion of a tree.) An alternative reason for the collapse at Venice was put forward by Alban Caroe (1949), although he gives no detailed explanation to support his hypothesis. In a brief chapter on belfries, bell-frames and bells in his book on old churches, he makes the statement: 'From time to time a request is put forward that a urinal shall be arranged in a medieval church tower. Many of the dangers which must be guarded against in any such provision are obvious, but it is worth remembering that it was insanitary use of this kind which caused the collapse of the great campanile of St Mark at Venice.' This paragraph, quoted in its entirety, is not further expanded by Caroe.

6.2 Collapses at Beauvais, 1284 and 1573

The enormous dimensions of Beauvais Cathedral are not in fact much greater than those of the great churches of the first half of the thirteenth century. The centre-line width between main piers of the choir of Beauvais is 15 m, almost exactly that of Bourges, Chartres, Amiens and Cologne (started 1248, but largely completed in the nineteenth century), and slightly more than Reims; the total width of the choir (about 42 m) is about the same as Bourges and less than all the others. Only the height of the vault, 48 m, is greater than the others, and Cologne has a height of 46 m. As for the spacing of the piers in the axial direction, the three original bays of the choir at Beauvais varied from about 8 m to 9 m (extra piers were inserted after the collapse of 1284), and these bays were in fact slightly smaller than the largest bay at Amiens, and almost exactly the same as at Reims and Cologne.

Beauvais was started in 1225, and a first campaign of 20 years took the work up to triforium level. The main campaign, 1250–72, saw the choir, crossing and transepts completed, with the high vaults standing 48 m from the ground. In 1284 these vaults fell.

Contemporary records do not refer to any natural catastrophe, such as an earthquake or violent storm. The fact is that the cathedral had stood comfortably for 12 years, which is ample experimental proof that equilibrium had been achieved between thrust and counterthrust; a theoretical structural analysis confirms this stability, but this is really unnecessary. However, the collapse of 1284 is just about within the period required for consolidation of the soil, and perhaps the most likely cause of failure must be sought in some adverse geotechnical effect – after 700 years it is improbable that any precise mechanism can be found.

Alternatively, if some slow-moving action could be discovered which might lead to failure of an important structural component, and if that failure could somehow trigger an overall catastrophic collapse, then perhaps a cause other than uneven settlement might have been operating at Beauvais. Viollet-le-Duc postulated such an alternative cause, based on the fact that medieval mortar was slow to set, and liable to shrinkage over years or decades. His explanation may or may not be correct, but it is of the right kind, since it leads to a small failure with very large consequences.

Figure 6.2 shows Benouville's reconstruction of the state of Beauvais in the period 1272–84, and features discussed in previous chapters are evident. The two tiers of flying buttresses are required to span two side aisles, and they do this elegantly and economically with the help of an intermediate pier. In fig. 6.3 Viollet-le-Duc gives details of the construction; the section here is at the chevet, where the external buttresses are

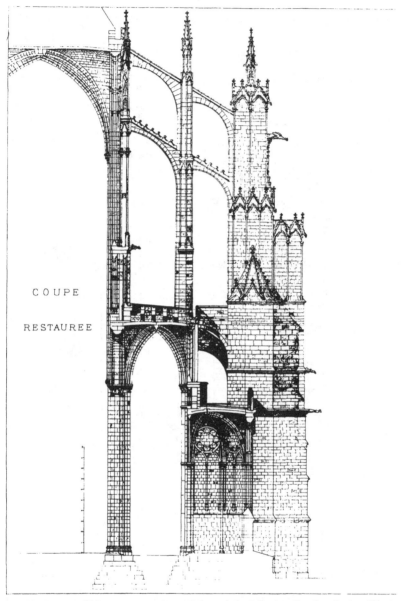

COUPE

RESTAUREE

Fig. 6.2. Benouville's reconstruction of the original design for the choir of Beauvais (from Benouville 1891–2).

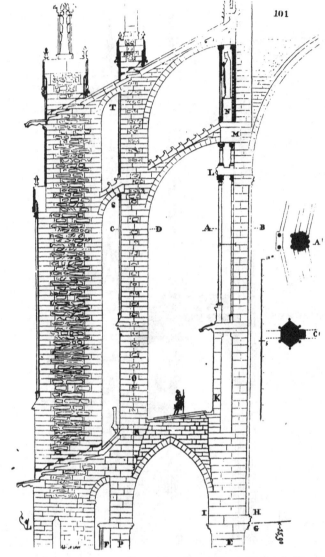

Fig. 6.3. Beauvais; part cross-section at the chevet (from Viollet-le-Duc 1858–68).

placed closer to the main fabric. Figure 6.4 gives a perspective sketch of the heads of the flying buttresses.

Viollet-le-Duc considers that the slender twin columns A failed (fig. 6.3). The mortar, slowly shrinking in the adjacent pier B, ensured that more

Fig. 6.4. Sketch of the heads of the flying buttresses of fig. 6.3 (from Viollet-le-Duc 1858–68).

and more load was thrown onto the twin columns until they eventually fractured. As a consequence the lintel L broke, and the great block M, effectively the tas de charge, loaded by the gigantic statue N, was no longer supported. Viollet-le-Duc then considers that gross deformation occurred, and that block M slid out, pushing aside the flying buttress. The vault thrust would then be unsupported, leading to collapse of the vault in that bay, almost certainly completely across the church. The collapse would spread, since adjacent vaults to the east and west would not be supported.

Details of this explanation may of course give rise to argument, but the broad mechanism (as an alternative to a geotechnical mechanism) is certainly plausible. Buckling of slender monolithic columns adjacent to stone and mortar masonry can certainly be seen elsewhere – as a single example, the sixteen Purbeck marble shafts surrounding the central pier of the Chapter House at Wells are in some cases markedly displaced (but these shafts are 'decorative', and their failure does not impair the integrity of the whole structure).

Repairs were made to Beauvais over the next 50 years, and the choir had been rebuilt by about 1337 before the work was interrupted by the Hundred Years War and the English occupation. It was not until 1500 that a renewed start was made on the transepts, and, with choir and transepts complete, but still with no nave, a crossing tower was begun in 1564. This tower (fig. 6.5), was completed in 1569, and rose 153 m from the ground; it alarmed the Chapter from the first.

The Venice Campanile is only 84 m high, and it differs in a major respect from the tower at Beauvais, and from any crossing tower. The walls of the Campanile spring directly from the foundations, whereas a tower at a crossing is carried on four isolated piers; in the case of Beauvais, the crossing piers rise at least 16 m before they receive any semblance of support from the arcading of the aisles. Thus a mass of thousands of tonnes of masonry depends for its stability on the good construction of the four piers, and these in turn rely on support from foundations that a modern engineer would consider grossly overloaded. At Beauvais, settlements inevitably occurred.

The Chapter sought advice, and two King's masons came from Paris to examine the work about two years after completion of the tower. They found the four main crossing piers were beginning to lean. Those on the choir side, to the east, were out of plumb by 1 in. and 2 in., but were not thought to be dangerous, since they were considered to be well-constructed throughout their thicknesses. The principal danger lay in the other two piers 'tirant au vide', that is, next to the non-existent nave, which were out of plumb by 5 to 6 in. and by 11 in., for lack of 'contreboutement'.

The King's masons proposed as remedy the immediate construction of two nave bays, and the strengthening of the pier foundations. In the meantime, temporary walls were recommended between the crossing piers (cf. the strainer archers at Wells, installed over two centuries earlier).

Fig. 6.5. Beauvais Cathedral, 1569 (frontispiece from Desjardins 1865).

The Chapter procrastinated, seeking further advice, and it was only two years later that the work was put in hand, on 17 April 1573; on 30 April, Ascension Day, the tower fell. All accounts agree that the two 'open' crossing piers failed first.

The clergy and people had just left the cathedral in procession; only three people were left inside the church, and all three escaped. Legends have grown up round this spectacular collapse. For example, no one would undertake the dangerous job of clearing the rubble in the partly destroyed building. Finally, after four months, a condemned criminal was offered his life if he would demolish the ruins. He had only just started when his footing gave way, and he fell, but managed to catch hold of a rope hanging from the roof beams, and so climbed to safety. 'The rope destined for the hangman's noose proved the salvation of this wretch.'

By 1578 all masonry repairs had been made, but the tower had not been rebuilt, and all money set aside for the nave had been spent. In 1605 the decision was taken to do no more, and Beauvais became what it is today, a choir and transepts with no nave and no tower.

6.3 The crossing tower

It is clearly necessary to provide lateral support to a tower carried on four separate piers. If a crossing tower is considered in isolation for the moment, as a free-standing campanile perhaps, then the four piers will be seen to carry arches which in turn carry the bulk of the four walls of the tower. The weights of the walls are concentrated by the arches into the piers, and so to the foundations, and, in this process, each arch will generate a sideways thrust in the usual way. Thus, for a symmetrical tower, each pier will be subjected at its head to an outward diagonal thrust. At Beauvais the masonry of the choir and transepts absorbed these thrusts to the north and south and to the east, but no buttressing had been provided to the west.

A satisfactory crossing tower, then, will be buttressed on all four sides by the fabric of the church, and the tower is in fact liable to damage that fabric during the first few years after construction. It was noted in Chapter 2 that the loads in the piers will apply very high pressures to the soil beneath the foundations, even if those foundations are built out underground in an attempt to reduce the stresses. As a result it is very common to observe that the tower at the crossing has settled differentially with respect to the rest of the fabric; a tower may 'punch through' the crossing by as much as 200 or 300 mm. In this process of settlement considerable distortion will be imposed on the abutting (and supporting) masonry; string courses may be seen to bend down in the choir, nave and transepts as the crossing is approached, and cracking of masonry may be observed one or perhaps two bays away from the crossing.

In general such defects will become apparent, and movement will virtually cease, within the soil-mechanics time scale of 20 years or so after completion of the work. During this process of consolidation equilibrium is established in the soil, the weight of the tower being supported both by direct contact of the soil particles and by water pressure. Should the water table be lowered subsequently, then renewed settlement may occur, and this may still take place many years after completion of the building. Almost certainly such settlement occurred at St Paul's in London towards the end of the last and in the first quarter of this century, as a result of the removal of water from the surrounding region. The 1921 Commission (quoted by Harvey 1925) estimated the mass of the dome structure supported by the eight great piers to exceed 6700 tonnes, and the effects of the settlement at the crossing may be seen clearly in the four abutting arms of the Cathedral (for example in the cracking evident about half way down the south transept).

These defects due to settlement are additional to those mentioned earlier in the chapter (fig. 6.1) – drift apart of the two skins of a wall will occur, with voids developing in the weak rubble core. The voids can be filled by injecting the walls with grout; this will close existing cracks, but will not provide a tensile connexion through the thickness of the wall, and drift will continue. As noted above, the inner skins of the four walls will press against each other and, unless they buckle, will effectively be locked in place. The grout will do little, however, to prevent the outer skins from continuing to move.

The traditional 'cure' for this problem, and a reasonably effective cure, is to insert ties across a tower at one or more levels. The ends of the ties are secured by iron (or stainless steel) pattress plates bearing on the external walls of the tower, and metal straps may also be used to distribute the loads from the ties. (In any case the local bearing areas of masonry should be consolidated with grout in order to reduce local stressing.) Such tying and strapping was, for example, employed by Scott in his repair work to the West Tower of Ely Cathedral in the 1860s, and was in principle entirely satisfactory. Some 100 years later, however, much of this ironwork was very badly corroded; the structural scheme had worked well, but intensive damage was being caused to the original stonework. All the external strapping to the West Tower was removed in

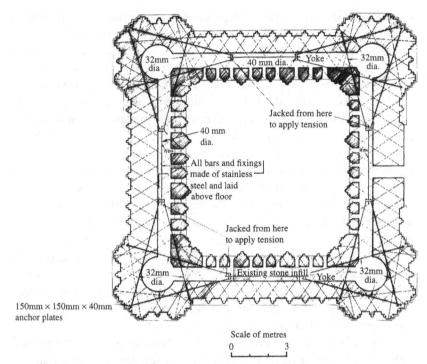

Fig. 6.6. Ely Cathedral: West Tower. Strengthening with stitches and ties.

the 1970s, and was replaced by a scheme of which one level is illustrated in fig. 6.6. This scheme echoes exactly that of Scott, but uses stainless steel instead of iron and borrows techniques from modern reinforced-concrete construction.

In effect, fig. 6.6 shows a complete 'reinforced-concrete' ring beam constructed within the fabric of the tower. A large number of horizontal holes were drilled diagonally from the outer faces, and reinforcing rods were inserted tying together the two skins and stopping only a centimetre or so from the faces of the masonry. Cement grout was then injected under low pressure into all these holes so that, effectively, a complete ring beam was established within the masonry. The holes were plugged with new stone flush with the ashlar so that no visual sign of the work can be detected.

In addition, as will be seen from fig. 6.6, massive ties were inserted connecting together the four corners of the tower, but, unlike those of Scott, not causing obstructions by passing through the stairs. The old

ties, like the external strapping, have been removed. The new ties really
need the ring beam to pull against; they require stiffened fabric to be
effective. On the other hand the ring beam would have been effective
(the scheme of fig. 6.6 was repeated at three levels in the height of the
West Tower of Ely) without the addition of ties. In the 1992/93 repair
of the southern transept to the West Tower, a single level of Scott's ties
was removed and replaced only by a stitched ring beam.

The essential requirement for the stabilization of a masonry wall
is indeed to connect together in a positive way the two skins of the
construction; if this can be assured, then the traditional provision of
external ties becomes of secondary importance. The insertion of a ring
beam (reinforced concrete or perhaps steel) in the inner face of a tower,
perhaps in an attempt to strengthen the tower at bell-frame level, may
very well not provide this necessary connexion through the thickness of
the walls. Such a partial strengthening may therefore not achieve the
desired effect.

There are particular problems at the top of a tower. The tower may be
surmounted by a spire – as will be seen in the next chapter, a masonry
spire thrusts horizontally, as well as vertically, and must be properly
supported at the level of the parapet to the tower. If there is no spire,
there will be a timber roof, perhaps pyramidal and perhaps untied across
its span, and, just as for the timber roof over a nave or choir, the lateral
forces must again be contained. Effects on the masonry at the top of
the tower depend on the structural details of the connexion between the
spire or timber roof and the walls, but it is common to see the effects
of a spread in dimensions. This is often betrayed by open joints in the
parapet masonry, and vertical cracking may also be seen in the walls. It
is, of course, at the top of a tower that metal ties are most often seen
and, properly inserted, they can completely control the development of
further movement. Indeed, it is usually simpler to instal a set of tie bars
in the tower than to try to tie the timber roof (or spire) so that it imposes
purely vertical loads on the tower masonry.

6.4 The effects of bell-ringing

The forces generated by the ringing of bells depend, of course, on the
weights of those bells; they depend also upon the arrangement of bells
within the bell-frame, and the sequence in which the bells are rung. The
effects of the forces on the tower depend on the height at which the bell-
frame is installed, and the behaviour of a particular tower will be affected

by its relation to the rest of the fabric of the church. It is common, for example, to find a ring of bells in the west tower of a church, and that west tower may have a large window in its western face and a great arch into the church to the east; the north and south walls may consist of solid masonry. Such a tower will be stiffer in the east–west than in the north–south direction, and account will have been taken of this in the design of the bell-frame; if possible, more bells of the ring, and the larger bells, will be hung to ring east–west rather than north–south.

Despite such large and particular variations, it is possible to make some general observations. The Handbook (Frost 1990) gives much information, and some of the mechanics of bell-ringing has been covered by Ranald Clouston (1970) and by Heyman and Threlfall (1976). During one revolution of a bell, considerable forces are induced at the axis of rotation, and those forces must be reacted first by the bell-frame (traditionally of massive timber construction, and more recently of iron or steel), and then by the tower itself. The frame must be firmly attached to the tower to avoid 'battering', and it is for this reason that the tower may be strengthened with a ring beam in order to make firm connexions with the frame.

The values of the forces engendered by a swinging bell depend not only on its weight, but on the way in which the bearings are disposed in relation to the centre of gravity of the bell. Actual values may be found for a given ring of bells by performing two simple tests – attaching a weight to a bell rope and observing the angle of displacement, and timing the period of small oscillations in the mouth-down position. However, conventional values based only on the nominal weight of the bell suffice for most purposes.

The vertical force on the bearings, in addition to the dead weight of the bell, may be taken as a maximum of three times that weight during one revolution; thus a total vertical force of four times the weight of a bell will be produced during ringing. This has consequences for the behaviour of the bell-frame, but almost none for the behaviour of the tower itself.

The lateral force, however, taken nominally as a maximum of twice the weight of the bell, can have marked effects on the tower. (The actual value of the lateral thrust may be less than twice or as great as three times the weight of the bell.) Figure 6.7 is based on an illustration to Clouston's discussion, and several points may be noted. The time taken for a complete revolution of the bell in this example is 2 seconds, and this is typical for most rings. Thus for a ring of eight bells ringing changes,

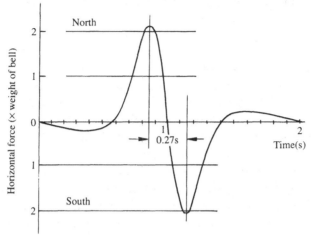

Fig. 6.7. Horizontal force at the bearing of a bell during one complete revolution (from Clouston 1970).

the bells will follow each other at intervals of about $\frac{1}{4}$ second. It will be seen from fig. 6.7 that the horizontal thrust, after a small initial dip, rises to a maximum of about twice the bell's weight, and then falls sharply to a maximum in the reverse direction, before dropping back to zero. The time between the two peaks in fig. 6.7 is almost exactly $\frac{1}{4}$ second. Thus if two bells of roughly the same weight, and swinging in the same direction (e.g. N–S), follow each other in sequence, then the horizontal forces will tend partly to cancel. This assumes, however, that both bells are ringing clockwise (or anticlockwise); if one rings clockwise and the other anticlockwise, then two peaks in horizontal thrust will be additive, and a doubled force will act on the tower.

When eight bells are ringing changes, in a sequence which is continually altering (and there are 40 320 possible combinations) there will be occasions when the forces from several bells will be additive and other occasions on which they cancel out. The arrangement of bells within a frame is a highly complex matter, both in the orientation (i.e. N–S or E–W), and in the roping which imposes clockwise or anticlockwise rotation. The fact is that, however the bells are installed, there is the possibility of large horizontal forces being generated, randomly or rhythmically, at time intervals of $\frac{1}{4}, \frac{1}{2}, \frac{3}{4}$ second, and so on at $\frac{1}{4}$ second modules. Moreover, the natural period of oscillation of a typical tower might be $\frac{1}{2}$ second, or longer for a massive tower, so that the horizontal pulses may well coincide with this natural period.

A first consequence for a tower is that it will be felt to move when the bells are rung. The movement may be small – in fact, *is* small – but the body is a sensitive indicator. The reason that vibrations are small has to do with the way that bells are rung. Although a bell at the top of its stroke can be just 'over the top', and kept in equilibrium by its stay and slider (it is dangerous to leave bells in this state, since an accidental tug on a bell rope can start a bell ringing), a bell-ringer does not make use of this equilibrium when ringing changes. Instead, the bell is virtually balanced, in temporary but unstable equilibrium, at the beginning and end of each stroke, so that is ready instantly to take its ever-changing place in a particular round. If the tower, and hence the bell-frame, are themselves moving, then achievement of this unstable equilibrium is not easy, and it is usually thought that a movement of $\frac{1}{16}$ in. (i.e. 1.5 mm) of the frame begins to affect the 'go' of the bells. If the displacement is double this value, 3 mm, then bell-ringing becomes difficult, and it may be concluded that, if bells are being rung at all in a tower, then the displacement at the level of the bell-frame will be less than say 3 mm.

An oscillation of 3 mm may be thought trivial, but in fact it has consequences for a tower considered as a vertical cantilever rooted in the ground. Using typical dimensions, simple structural theory shows that bending stresses of the order of 1 N/mm^2 are induced by the motion; that is, the stresses are at the same level as those induced by the dead weight of the whole tower. Moreover, the bending stresses can theoretically be both positive and negative – that is, tensile and compressive – so that dead weight compressive stresses may tend to be reduced or even reversed. Fortunately, the absolute magnitudes of about 1 N/mm^2 are low, so that any weak tensile resistance of the masonry may be mobilized to resist cracking.

However, another consequence of tower vibration – ringing cracks – can be observed in many towers housing bells. Some cracking is local, and can be attributed to poor installation of the bell-frame or of its supporting joists, but the vibration of a tower as a whole can cause long vertical cracks to develop at or near the centres of the four faces. If, for example, the tower is moving in the north–south direction, then the east and west faces, near their centres, will be subjected to potentially large shear stresses. As has been remarked, a material weak in tension is also weak in shear, and this is reflected in the development of the consequent cracks, which may often be seen passing through the heads of windows in the walls. The presence of the openings indeed weakens the walls in their ability to carry shear, but such cracks can develop also in solid parts of the masonry.

Cracks of this sort of small width are not really harmful, and the traditional way of dealing with them, by stitching, is effective. A structural stitch placed across a ringing crack will restore connexion between the separated portions of the tower, and will help to transmit the shear forces engendered by the vibrations which inevitably accompany bell-ringing.

6.5 Wind loads

A tower of 2000 tonnes will not be affected in the large by winds. A typical gust pressure is 1 kN/m², leading to a force of 24 tonnes on the face of a tower 30 m high with a side of 8 m. Clearly there is an enormous reserve against overturning of the tower.

On the other hand a force of 24 tonnes is greater than the forces generated by an average ring of bells; very few tenor bells exist with weights over 2 tonnes. Even the largest hand-rung tenor, that of the Anglian Cathedral in Liverpool, has a mass of not more than 4170 kg, and will give rise therefore to horizontal forces of less than 10 tonnes. Even if all the bells in a ring are sounded together, it is unlikely that forces on the tower will be generated that exceed those arising from wind.

Indeed, wind forces can help to propagate the 'ringing cracks' attributed to the bells. Wind forces, although greater than the bell forces, are however, intermittent; 40 320 combinations of an eight-bell ring can be achieved in something over 3 hours, during which time the tenor bell will have sounded 5040 times. The tower will have been subjected to a force of say 2 tonnes every 2 seconds for the period of 3 hours; the effects must be compared with those of an occasional wind gust of 20 tonnes. Both wind and bells lead in fact to cumulative damage to the fabric of the tower, but the process is slow for a well-built tower and a properly installed bell-frame.

Small details of a tower *are* affected by wind; for example, a pinnacle of say roughly square cross-section will blow over in a gale if its height is too great and it is not properly secured by dowels. The calculation is straightforward for a pinnacle of square cross-section $d \times d$ and height h, made of stone of unit weight ρ and subjected to a wind pressure p; the pinnacle will blow over if

$$h > \frac{\rho}{p}d^2.$$

Thus, for $\rho = 20$ kN/m^3 and $p = 1$ kN/m^2, the above formula gives $h > 20d^2$, where d and h are measured in metres. A pinnacle 0.3 × 0.3 m will blow over when its height exceeds 1.8 m.

Strangely, a conical pinnacle behaves differently; the tip of a masonry spire is examined in the next chapter.

7

Spires

A masonry spire is usually octangular in cross-section; the square tower is converted by squinch arches or other means near its top into a regular octagon, from which the spire springs. Such a spire surmounts the church of St Mary at Hemingbrough (Yorkshire, East Riding), which is surprisingly large for a small village; it was in fact a collegiate church under the Prior and monks of Durham. The height of the thirteenth-century central tower, about 18.5 m, is in keeping with the general mass of masonry, but the total height to the top of the spire is 54.4 m (fig. 7.1). The spire itself was added in the second quarter of the fourteenth century, and springs from slightly below the parapets of the tower; it measures 37.5 m, and thus forms two-thirds of the total height of the church. The visual impact is curious, although Pevsner believes that the composition 'happily breaks all rules of harmonious proportion'.

The 'diameter' of Hemingbrough spire at its base measures 5.50 m. A diameter is rather an imprecise measure for a spire whose horizontal cross-section is an octagon, but, as will be seen, some insight into the behaviour of spires can be obtained by treating a spire as a right circular cone, at least in the first instance. On this basis, Hemingbrough spire has a half angle α given by $\tan\alpha = 0.0733$; this is a half angle of just over $4°$ – that is, the full angle at the tip of the spire in the sketch of fig. 7.1 is just under $8\frac{1}{2}°$.

The thickness of the masonry at Hemingbrough is nominally about 8 in. or 200 mm. The ratio of diameter of the spire (at the base) to thickness is therefore 27.5. Ungewitter states as an empirical rule that this ratio should lie between 24 and 30 for a spire made of weak stone, but that the use of good quality stone allows a ratio of 30 to 36. (These numbers are examined further below.) If a spire is of constant thickness throughout its height then the ratio will of course decrease as the tip is

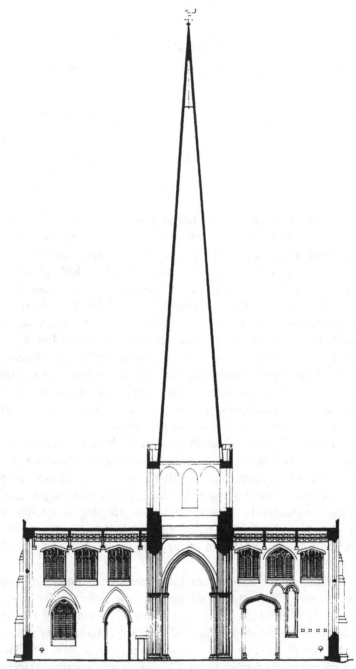

Fig. 7.1. Hemingbrough, St Mary. Cross-section looking east, from a drawing by Peter W Marshall FRIBA.

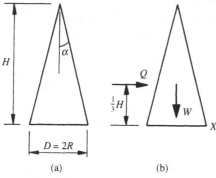

Fig. 7.2. A right circular cone as a model of the tip of a spire.

approached; indeed, the tip of a spire is, of necessity, solid. This solid tip is potentially vulnerable to the effects of the wind.

7.1 The spire tip

The modelling of a spire by a right circular cone reveals the stability problem at the tip. Figure 7.2 shows a length of tip H acted upon by its self-weight W and by a wind load Q. Now the self-weight is proportional to the volume of stone, and hence to the cube of the scale of the diagram, while the wind load is proportional to the presented area, or to the square of the scale; a long enough length of tip will therefore have sufficient weight not to be overturned in a gale. The critical length H below which there will be problems can be estimated by writing expressions for the overturning and restoring couples.

The weight of the cone in fig. 7.2 is

$$W = \frac{1}{3}\rho\pi R^2 H \tag{7.1}$$

where ρ is, as usual, the unit weight of material. The wind force Q arising from a unit wind pressure q has value

$$Q = qRH. \tag{7.2}$$

In equation (7.2) there should be in addition some factor reflecting the fact that the spire is octagonal and not right circular, that crockets at the quoins will upset the air flow and so on; although the factor cannot be known except for a particular spire, its value is somewhere near unity, and does not affect the present argument. Similarly the height $\frac{1}{3}H$

marked in fig. 7.2(b) is only approximate, and will depend on the actual details of construction. Nevertheless, the couple tending to overturn the cone about the point X is of the form $\frac{1}{3}HQ$, while the restoring couple is RW; if the cone is to be stable, then

$$R\left(\frac{1}{3}\rho\pi R^2 H\right) > \frac{1}{3}H(qRH). \qquad (7.3)$$

Using the relation $R = H\tan\alpha$, inequality (7.3) becomes

$$H > \frac{q}{\rho\pi\tan^2\alpha} = H_o, \text{ say.} \qquad (7.4)$$

For $q = 1$ kN/m² and $\rho = 20$ kN/m³, equation (7.4) for the Hemingbrough spire (for which $\tan\alpha = 0.0733$) gives $H_o = 2.96$ m. It was noted that a factor of order unity was omitted from equation (7.2); nevertheless, these calculations indicate that a right circular cone of tip angle $8\frac{1}{2}°$ will blow over if its length is less than about 3 m.

In fact, the appropriate unit wind pressure at Hemingbrough is greater than 1 kN/m², and the unit weight of the masonry may be less than that assumed, so that a length even greater than 3 m will be necessary for stability. The practical conclusions are clear – the top few metres of a masonry spire must be built in such a way that the courses of stone are tied together, and so able to resist the tensile forces generated by the action of wind. In practice, courses will be connected with metal dowels, usually made now of stainless steel.

Hemingbrough spire is exceptionally slender, with its half angle of about 4°, and the critical length of the tip is correspondingly large. Using the same values for wind load and self-weight, the critical lengths for spires with half angles of 5°, 6° and 7.5° are reduced to 2.08, 1.44 and 0.92 m respectively. There are no great constructional difficulties with such shorter lengths, but it is a matter of observation that the tips of existing spires, when examined through field-glasses, either show some distress in the way of loose mortar joints or other defects, or betray by the presence of new stone that they have been rebuilt.

There is a special device which is sometimes used in practice that confers stability on the spire tip. Spires commonly have wind vanes, and the vane rod on the axis must be continued, for anchorage, for some distance down the spire, and indeed it helps to reinforce the masonry courses. This rod may be lengthened, and a weight attached to its bottom, freely suspended within the hollow of the spire. Effectively a weight W' will be added to the weight W in fig. 7.2(b), and inequality (7.3) necessary for stability is replaced by

$$R(W + W') > \frac{1}{3}HQ, \tag{7.5}$$

which leads to

$$W' > \frac{1}{3}qH^2 - \frac{1}{3}\rho\pi \tan^2 \alpha H^3. \tag{7.6}$$

The right-hand side of inequality (7.6) takes a maximum value for $H = \frac{2}{3}H_o$, and, using this value,

$$W' > \frac{4}{81}qH_o^2, \tag{7.7}$$

where H_o is the critical length of tip previously calculated and given by (7.4) above. If the weight W' given by (7.7) were hung from the vane rod, then no tensile forces would be induced by the wind anywhere in the spire. No 'factor of safety' has been introduced in the above argument; if a factor of 2 were used, then a design rule for the added weight would give double the value implied by (7.7), or say

$$W' = \frac{1}{10}qH_o^2. \tag{7.8}$$

Thus for Hemingbrough, for which H_o was determined as 2.96 m for $q = 1$ kN/m^2, $W' = 0.88$ kN (i.e. a 90 kg mass). (For a more realistic wind loading for Yorkshire q might be as high as 1.5 kN/m^2, leading to $H_o = 4.44$ m, and $W' = 2.96$ kN, i.e. a 300 kg mass attached to the vane rod.)

7.2 The spire as a conical shell

Calculations of wind forces, similar to those for the tip, may be made for the hollow spire. Continuing for the moment with the octagonal shell modelled as a circular cone, the weight of a reasonably long length H of spire having uniform wall thickness t is given closely by

$$W = \rho\pi H^2 t \tan \alpha. \tag{7.9}$$

This weight is proportional to the square of the height H (and not to the cube of the dimensions, as implied for the weight of the tip, equation (7.1)). The wind force, however is exactly as before, so that, for stability,

$$R\left(\rho\pi H^2 t \tan \alpha\right) > \frac{1}{3}H\left(qRH\right), \tag{7.10}$$

or

$$t > \frac{q}{3\pi\rho\tan\alpha}. \tag{7.11}$$

A certain minimum weight of spire, that is, a certain minimum thickness t given by inequality (7.10), is necessary for stability, and this thickness is independent of the actual height H of the spire.

Inequality (7.11) may be written more conveniently by using the critical length H_o from equation (7.4); substitution gives

$$t > \left(\frac{1}{3}\tan\alpha\right)H_o. \tag{7.12}$$

Thus for Hemingbrough, for which H_o was found to be 2960 mm for a unit wind pressure of 1 kN/m^2, the minimum value of t is 72 mm. Corresponding values for half angles α of 5°, 6° and 7.5° (as before) give minimum values of t of 61, 50 and 40 mm respectively.

No factor of safety has been included in the calculations leading up to inequality (7.12); a factor of 2 on the above minimum thicknesses would lead to the formula

$$\frac{2H_o}{t} < 3\cot\alpha. \tag{7.13}$$

Thicknesses calculated from this formula still give values which are in general markedly less than those found in practice. It may be concluded that thin-walled spires of usual dimensions are not in danger of overturning under the action of wind.

Stresses in a conical shell may be established by using simple membrane theory, with equations analagous to those of Chapters 3 and 4 for domes and vaults. It seems evident that self-weight alone produces purely compressive stresses in the spire, and indeed the weight is supported by longitudinal compressions along the generators and by 'hoop' compressions at every horizontal section, as suggested in fig. 7.3. The value of the compressive stress along the generators at the base of the spire is given by

$$\sigma = \frac{1}{2}\rho H \sec^2\alpha \tag{7.14}$$

or, very closely, by $\frac{1}{2}\rho H$, since $\sec\alpha$ is almost unity. Thus a spire of height 30 m and of material having unit weight 20 kN/m^2 will have longitudinal compressive stresses at the base of 0.30 N/mm^2; as usual this (low) value is independent of the thickness of the spire. The value of the hoop stress will be only a small percentage of the longitudinal stress.

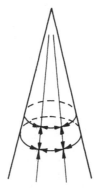

Fig. 7.3. Stresses in a conical spire due to self-weight — longitudinal and hoop compressions.

Wind forces, however, will produce additional compressions on the lee-ward side and tensions on the windward side of the spire, and these must be superimposed on the basic self-weight compressive stresses. Again, membrane shell theory may be used to estimate the value of the wind stresses, although simpler theory is in fact adequate; net tensions start to develop when the wind forces are about half those necessary to overturn the spire. Thus very slender spires, such as that of Hemingbrough, might move locally just into the tensile regime under the action of maximum wind; less slender spires will stay in compression.

7.3 The octagonal spire

All of the discussion on the behaviour of a right circular cone may be applied to the real octagonal spire, provided that spire has a minimum thickness of construction. A circle of diameter D can be contained within two concentric regular octagons separated by a distance t if $D \sin^2(\pi/16) < t$, that is, $D/t < 26$. Many spires satisfy this condition throughout their lengths (cf. Ungewitter's rule of 24 to 30), and in all these cases the results for the cone confirm the stability of the octagonal spire. On a cross-sectional plan of the spire a circle will be contained within the thickness of the octagonal walls; the cone may be imagined as embedded within the actual masonry, and, correspondingly, the stress solutions for the cone may be regarded as satisfactory equilibrium states for the actual spire.

Alternatively, the membrane forces sketched in fig. 7.3 for the smoothly

Fig. 7.4. Leading forces in an octagonal spire (cf. fig. 7.3).

turning conical surface might be 'lumped', for the octagonal spire, into inclined forces along the quoins together with discrete horizontal forces; in fig. 7.4 the spire has been replaced notionally by a skeletal structure with say 'square' panels (actually trapezoidal). Certainly overall equilibrium can be satisfied by the system sketched in fig. 7.4, and indeed the material within the skeletal trapezoids could be omitted, in which case no extra forces would arise.

Spires have been built with a large amount of material omitted; there are examples in Ulm, Regensburg, Cologne and elsewhere. Figure 7.5 shows the spire of Freiburg im Breisgau (first half of the fourteenth century). An immediate conclusion that might be drawn from a study of such spires is that accidental damage to a conventional solid-faced octagonal spire (a stone lost by impact, for example, or by decay, leaving a hole through the surface) is very unlikely to distress the spire as a whole.

Further, comparison of figs. 7.3 and 7.4 indicates that the idealized 'lumped' model, while being 'safe' by the master theorems, may also not be far from representing the actual behaviour of a spire. It would then remain to examine the behaviour of the infill panels in the real or imagined trapezoids. It seems evident that, if an infill panel were too thin, it would be in danger of collapsing inwards, whether it were conventionally solid and of uniform thickness, or 'decorative', as at Freiburg im Breisgau. Stability of a panel would also, intuitively, seem to depend on the slenderness of the spire; a very slender spire would have nearly vertical faces, which could be made thinner than those of stockier construction.

Fig. 7.5. The spire of Freiburg im Breisgau.

Fig. 7.6. A tilted masonry spire at such an angle that all stresses remain compressive.

The problem can be resolved by making some assumptions (some of them not clearly justifiable) and by incorporating a safety margin, to give an empirical rule:

$$\frac{D}{t} < 3 \cot \alpha, \tag{7.15}$$

where α is as usual the half angle of the spire. Thus for Hemingbrough a satisfactory value according to this rule would be $D/t = 41$; corresponding values for $\alpha = 5°$, $6°$ and $7.5°$ are 34, 29 and 23, (all to be compared with Ungewitter's range 24–30, or 30–36 for stronger stone).

The formula (7.15) may be compared with (7.13); the latter ensures stability against wind forces. Only for exceptionally slender spires in areas of high exposure would (7.13) be more critical.

7.4 Leaning spires

As a curiosity, which applies more to timber than to masonry construction, a spire may tilt by 50 per cent more than its half angle before tension is developed along a generator. Thus a spire of half angle $6°$ can lean by $9°$ (fig. 7.6) and still stay in a regime of compressive stress. This result is due to the low centre of gravity of a conical shell, and such a spire can in fact tilt twice as far before overturning.

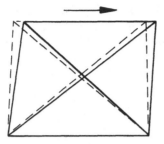

Fig. 7.7. One panel between quoins of a timber spire; shrinkage of one of the diagonal braces will allow the panel to deform.

7.5 Timber spires

The general approach of this chapter to the analysis of the masonry spire applies to some extent also to the timber spire. The framing of the timber spire is such that the system sketched in fig. 7.4 applies; timbers run along the eight quoins, and are connected to each other by horizontal members dividing the spire into a series of 'lifts'. The quoin timbers may run through two or more lifts, depending on the size of the spire. The skeletal framing is usually very flexible; it is the timber boarding forming the faces of the spire (and covered with lead or shingles) that confers rigidity on the whole construction. When complete, the timber spire may be viewed as a dual structure. On the one hand the skeleton of fig. 7.4 is maintained rigidly by the face boarding, and the skeleton may be thought to carry the loads; on the other, the skeleton may be thought of as permanent formwork for the effective shell structure of the eight faces. (The stone spire may of course also be viewed in this dual way.)

Timber shrinks after it has been assembled, and it is uneven shrinkage which causes a timber spire to tilt. It is also shrinkage which causes a spire to twist. (Viollet-le-Duc, in his article *Flèche*, is clear that timber spires are always subject to failure by torsion.) Many timber members have been omitted from the skeleton implied by the sketch of fig. 7.4; in particular, each of the trapezoidal panels will be diagonally braced as indicated in fig. 7.7. Without such braces the panel would only be restrained against deformation by the stiffness of the joints at the corners, and of course by the final boarding. However, the timber braces will in practice shrink, and movement of the panel will take place as indicated until one of the braces becomes tight. It will be appreciated that the panel sketched in fig. 7.7 is one of eight in the lift; distortion of one panel will induce a similar movement in the adjacent panel, and so round a complete circuit of the octagon. Moreover, if the quoin timbers run through to the lift above, the next lift will also twist.

Horizontal boarding will offer little resistance to movement of this sort. Diagonal boarding, however, will effectively provide continuous cross-bracing to the panels.

8

Some Historical Notes

8.1 Villard de Honnecourt(1)

The sketchbook of Villard de Honnecourt (*c.* 1235 and later) gives the lie once and for all to any idea that cathedral building was an amateur occupation. The book is written for skilled professionals. Figure 8.1, for example, shows one of the pages. The centre caption at the bottom refers to the survey under way: Pa(r) chu p'ntom le hautece done toor – *Par ce moyen on prend la hauteur d'une tour* (how to take the height of a tower). The standards of lining, levelling and plumbing employed in the construction of Gothic cathedrals were outstanding. On the right of the figure, an arcade is being set out – how to set up two piers at the same height without plumbline or level. On the left one can find a big medieval joke: Par chu tail om vosure pendant – *Par ce moyen on taille une voussure pendante* (how to construct a hanging voussoir). Villard was a master. Why then has he had no public recognition? Why are no streets, or satellites, named after him? The answer is that Villard was in fact a minor architect, like Vitruvius. The only thing that they have in common is that some of their precious manuscripts accidentally survived; nothing is otherwise known of any of their work. Even Villard's written legacy has suffered attrition; in the fifteenth century there were forty-one leaves, recto and verso; now only thirty-three remain. What is clear (and is argued beyond all doubt by Paul Frankl) is the firm thread that ties Villard to Vitruvius – explicitly, one finds the 'toys' (the eagle that turns its head to the deacon when he reads the Gospel) and, implicitly, the numerical rules of proportion.

Villard deals with topics other than *maçonnerie*. He sketches timber roofs, shows examples of the 'engines' used for his trade, and exposes geometrical rules of construction (as does Vitruvius). A large number

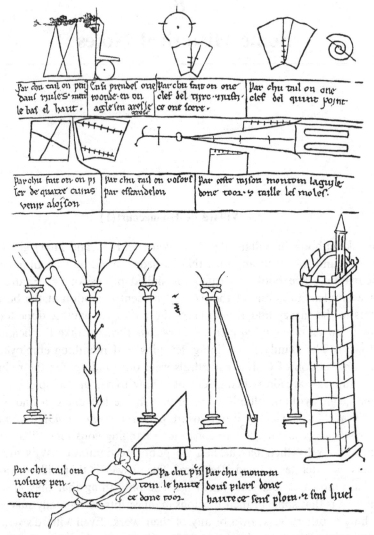

Fig. 8.1. Page from the sketchbook of Villard de Honnecourt (from Willis 1859).

of plates are devoted to portraiture, with Gothic faces and intensely sculptural drafting of drapery. The lions that appear in these sketches are fanciful – 'drawn from life', says Villard, although it is obvious that no real lion had actually served as model. Birds, dogs, horses, and ostriches also abound. Villard may not have seen any of these animals, but he had journeyed – to Hungary, for instance, a great voyage for a country lad who had been apprenticed in Picardy – and on these journeys he recorded, for his lodge, the new and extraordinary inventions of the 'golden age' of Gothic, which lasted for a century and a half, from 1140 (the abbey church of St Denis) to 1284 (the collapse of Beauvais).

8.2 Before Villard

The thesis of this book has been that the problem of the design of masonry is essentially one of geometry. The calculation of stress is of secondary interest; it is the shape of the structure that governs its stability. All surviving ancient and medieval writings on building are precisely concerned with geometry; from the viewpoint of the modern structural engineer, the ancient and medieval rules were essentially correct. Architects working to these rules had, no doubt, an intuitive understanding of forces and resulting stresses, but this would not have taken a form that would be of use in design, and there is no trace in the records, over the two or three millennia for which they exist, of any ideas of this sort.

Instead, it is likely that the design process would have proceeded by trial and error, by recording past experience and venturing, more or less timidly, into the unknown. Models were also used. A large-scale model would serve many functions – to demonstrate the design to the commissioner, to solve problems of stereotomy and, finally, to prove the stability of the finished full-scale structure.

The recording of past experience can be done verbally or by drawing. Chapters 40, 41 and 42 of Ezekiel, for example, record at interminable length the sizes of gateways, courts, vestibules, cells, pilasters and so on, for a great temple; part of a building manual of about 600 BC seems to be bound in with the books of the Old Testament. Of great interest is Ezekiel 40:3 and 5: '...I saw a man...holding a cord of linen thread and a measuring-rod...The length of the rod...was six cubits, reckoning by the long cubit which was one cubit and a hand's breadth.'

The dimensions which are given in the manual are in cubits and palms. The Hebrew cubit (about 17.7 inches) was divided into 6 palms; the 'royal' cubit – a cubit and a palm – was therefore about 20.7 inches, very close to the Greek standard of 7 palms. What the man was holding in Ezekiel was the 'great measure', without which work could not proceed on an ancient or medieval building site. This particular measure was 6 cubits in length, no doubt marked with a sub-division of palms, and could therefore be used to establish the major dimensions of rooms as well as small individual dimensions, merely by using the numbers listed so diligently in the books of Ezekiel. Once such numbers had been recorded, either in a manual or on a drawing, then they could be transferred to the site when the great measure had been physically constructed.

The essential feature of the great measure was that it was a part of the building. It was not an absolute 'yardstick'; if it were initially cut slightly smaller, then a slightly smaller building would result from the same building plan. All of this is apparent in Vitruvius, writing five centuries later, in the first of his six main concepts of the theory of architecture, *ordinatio*. Frankl's exegesis makes it clear that the *ordinatio* is nothing other than the great measure itself; *ordinatio*, says Vitruvius, is made up from *quantitas*, and *quantitas* are modules taken from the building. To make this clear, Frankl uses the example of a sculptor creating a human figure. The figure can be made to any scale but, whatever size is chosen, the ratio of one part of the sculpture (e.g. the head) to any other part (e.g. the hand) will be the same. Once the dimension of a component of the statue has been fixed – the foot, for example – all other components can be expressed in terms of that foot, and this unit of measure is the module, or *quantitas*.

All of this seems unnecessarily complicated today, when every worker on a building site has a knowledge of decimals, and has a standard folding metre rule, subdivided into centimetres and millimetres. Greek, Roman and medieval buildings were, however, constructed without any such standard yardstick, but with a local great measure established at the start of the project. Ground plans could be laid out with this great measure, and all components of the building – the height of a column, the width of a column, the space between columns – were expressed in terms of the modules which made up the *ordinatio*.

There then arose a problem, intellectually fascinating but, for those versed in the decimal notation, of no practical consequence whatsoever, concerned with the laying-out of dimensions which could not be expressed in terms of the modules. The rational numbers can be measured by sub-divisions of the module but, no matter how finely the module is sub-divided, the irrational numbers, which appear to have obsessed Greek mathematicians, cannot be so measured. Vitruvius was explicitly aware that the square root of 2 is irrational, and cannot be expressed as a 'number' on a great measure. However, he was also aware that this dimension can be constructed, and he gives this construction for doubling the square, to be followed immediately by a discussion of the Pythagorean 3:4:5 triangle.

8.3 Villard(2)

It is this preoccupation with numbers which is at the heart of the recorded 'secrets' of the medieval masons: the concerns of Vitruvius are the concerns of Villard. Villard's manuscript is infuriating for the modern reader; it was written for those who already knew the fundamentals of Gothic design, and is therefore silent on matters of fundamental interest. The drawings and text are concerned, on the one hand, with the recording of new or interesting buildings (some seen by Villard on his travels) and, on the other, with problems of geometrical construction.

The ground plans for great churches are of startling modernity: they can be read instantly by a present-day architect. Figure 8.2, for example, shows the plans of an existing church (Saint-Étienne, Meaux, at the bottom of the figure). The plan at the top is a new design, devised by Villard and a visiting master, Pierre de Corbie, *inter se disputando*; it shows the complex and ingenious vaulting arrangements for a double ambulatory round a choir with a circular termination.

Elevations, on the other hand, seem to be hopelessly out of scale, giving, for example, exaggerated prominence to particular architectural features. They are like those contemporary sketches in which a hand holds a stylized magnifying glass to enlarge one section of the work. This is precisely Villard's aim, and the medieval architect would have had no difficulty with these elevations. Once the ground plan has been fixed, and a set of numbers established by which, in proportion, the dimensions of every member of the church can be constructed, then there is no ambiguity about any of the vertical detail. Figure 8.3 shows the plan of the tower at the cathedral of Laon, taken, as Villard says, 'at a level above the first windows', that is at the level just above mid-height of the (nineteenth-century) sketch in fig. 8.4. Villard's own elevation, fig. 8.5, may be compared, to show its exaggerated detail (and its lack of mastery of perspective). No one has yet given a satisfactory explanation for the hand.

8.4 The expertises at Milan

In Villard's treatise there is a wealth of practical problems of mensuration, both of surveying and of detailed stone-cutting (how to find the radius of a shaft when the centre is inaccessible, how to halve the area of a square). It is secrets such as these that must have been given to every

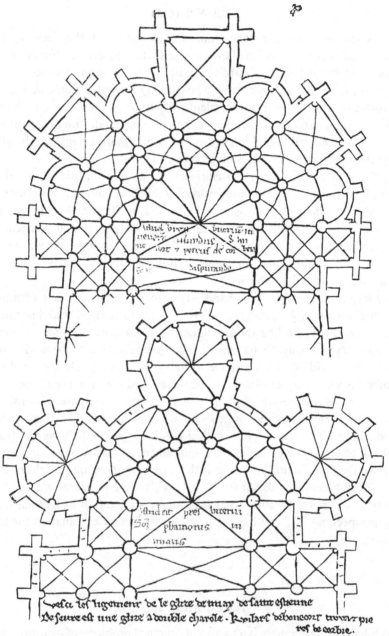

Fig. 8.2. Villard's plan of the church of Saint-Etienne, Meaux (from Willis 1859).

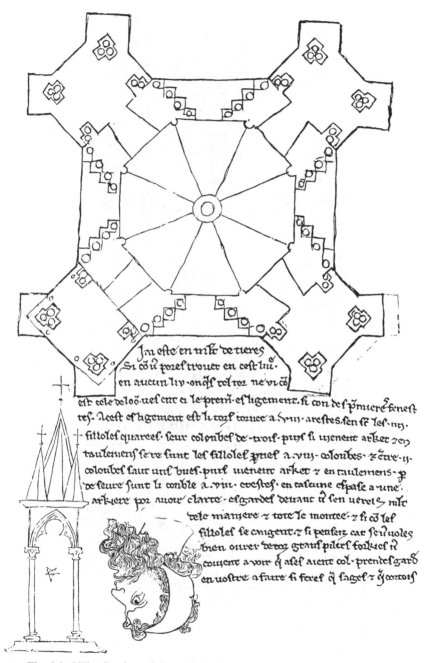

Fig. 8.3. Villard's plan of the cathedral tower at Laon (from Willis 1859).

Fig. 8.4. Nineteenth-century sketch of Laon (from Willis 1859).

Fig. 8.5. Villard's elevation of Laon (cf. fig.8.4) (from Willis 1859).

new member of a masonic lodge; without them, masons could not begin to learn their trade.

It was problems of mensuration that became acute in the building of Milan Cathedral. The cathedral was started in 1386, over a hundred years after the end of the High Gothic period, and difficulties of construction led to two well-documented expertises, in 1392 and 1400. The original design was *ad quadratum*; that is, the height of the work to the top of the high vault should be the same as the total width of the nave and four aisles – the transverse section was to be inscribed within a square. In 1391 the work had advanced to the point where the height of the piers had to be finalized, and doubts were expressed about the original intention. The Milan lodge sought advice from that of Cologne, but finally accepted recommendations from Stornaloco, a mathematician from Piacenza.

The clear (internal) width of Milan Cathedral is 96 braccia (the braccio, the Milanese 'cubit', is just under 2 feet, or about 0.6 m). Stornaloco proposed that the height should be 84 braccia, that is, that the construction should be *ad triangulum*, with the transverse section contained within an approximate equilateral triangle. It is precisely with this question of approximation that the Milan lodge needed advice from a mathematician. A true equilateral triange on a base of 96 braccia has an irrational height, not measurable by the great measure, of approximately 83.1 braccia. Stornaloco proposed that this figure should be rounded up to 84 braccia.

In summary, the great measure for the ground plan at Milan is of 8 braccia, and the designation *ad quadratum* means that the same great measure of 8 braccia was originally intended for the elevation. Stornaloco's proposal of 84 braccia not only eliminated the irrational square root of 3 but also, in effect, fixed a great measure of 7 braccia for the elevation. The proposal was disputed, and led to further discussion, involving Heinrich Parler of Ulm, who wished to return to the *ad quadratum* form; the Italian experts, on the other hand, wished to reduce the great measure for the elevation still further, to 6 braccia. In the event, they accepted Stornaloco's figure of 28 braccia for the height of the piers of the outer aisles (reduced in the event to 27.5 braccia to accord more exactly with the equilateral value of about 27.7), but above this level the work was completed to a vertical great measure of 6 braccia. As the horizontal great measure was unalterable at 8 braccia, the mensuration above the level of the piers was 'Pythagorean'.

All of this expertise was concerned, then, with establishing the *ordinatio* for the work, so that stones could be cut to fit exactly into the determined overall dimensions, and so that the dimensions of each component of the work could be laid out by reference to the great measures. Of great interest is the fact that one *ordinatio*, of 8 braccia, was used for horizontal dimensions, and another *ordinatio*, of 7 braccia for the lower work and 6 for the upper work, for vertical dimensions.

Building proceeded reasonably smoothly until 1399, when again a major dispute led to the holding of another enquiry. On this occasion, Giovanni Mignoto (actually Mignot) came from Paris, and Giacomo Cova from Bruges; they were joined within the next year by eight Italian architects to form a full-scale commission of enquiry. Mignot started, at the turn of the year 1399–1400, by drawing up a list of 54 points in which he found the work at Milan to be defective. The second half of the list consists essentially of trivia, but even the first half, of greater importance, is curiously arranged. Mixed up with objections which, if correct, are clearly of prime importance – that the buttressing was insufficient, for example – are objections of a different kind – that the canopies are set too high above the carved figures, or that the capitals and bases of the piers were not in their right proportions. These points are given equal prominence and are dealt with equally seriously by the Italian defenders of the work. Mignot was not satisfied with the responses, and it does seem that the Italians were inventing arguments to support their views, rather than appealing to some more rational and absolute base for discussion.

As an example, which particularly upset Mignot, the Italian assertion *...archi spiguti non dant impulzam contrafortibus* may be quoted. This occurred in a reply to Mignot's criticism that the buttressing was weak. The first statement was that the buttresses were well built of strong stone, connected with iron cramps, and well founded, so that they were indeed strong enough. Secondly, without prejudice to this statement, the buttresses were really not necessary, since pointed arches do not thrust (*archi spiguti ...* etc.). Finally, it had already been decided to tie the heads of the columns with strong iron ties to absorb any possible thrust.

There are grounds for saying that the thrust of a pointed arch is less than that of a round arch of the same span, but Mignot may perhaps be excused his bad temper. He certainly emerges as a much deeper scholar, versed in the theory of building, to the point where the Italians sulkily fell back on the statement *scientia est unum et ars est aliud*. Here *scientia* must not be coloured in translation by a modern view of science; *scientia* meant the theory embodied in the rules known by the master architect. Similarly *ars* contains no essential aesthetic element; it is the stonemasons's art, that is, the practice of construction, that is the intended meaning. In their statement that 'theory is one thing and practice another', the Italians were saying to Mignot that his theoretical rules were all very fine, but that they actually knew, in practice, how to build a cathedral.

Mignot's reply, *ars sine scientia nihil est* (practice is nothing without the theory), seems to herald the dawn of a new age of architecture. It was, in fact, nothing of the sort. Mignot had a book of rules with him which governed the design of great churches; this was his *scientia*. He had applied these rules to the work as he found it at Milan, and by these rules had found it wanting. He may have had a more comprehensive set of rules, or even better rules, than the Italians, but it seems clear from the records of the expertise that Mignot himself did not understand his own scholarship. His mixing together of 'aesthetic' and 'structural' criticisms implies that he did not, in any deep sense, understand either sort of rule – he merely knew that rules were being broken. Mignot's rule book had probably been compiled one or two centuries earlier, in the middle of the High Gothic period, and had lain in the lodge as a dead set of precepts whose purpose was increasingly dimly perceived.

The Italians had the final word. The Milan lodge doggedly rejected external advice on major problems of construction, and built the cathedral according to their own rules. The cathedral has survived for nearly 600 years.

8.5 The Renaissance

The reference in the expertise at Milan to *ars* and *scientia* echoes Vitruvius. In his first chapter, Vitruvius discusses the difference between *fabrica* and *ratiocinatio*, and stresses that both are necessary for the training of an architect. Vitruvius was read throughout the medieval period, and was copied again and again for use in monastic schools and in the masonic lodges. Mignot's Paris lodge and the Milan lodge would equally have produced architects trained in both practice and theory.

It is ironic that Vitruvius, remembered in the lodges and enshrined, in however altered a form, in the Gothic architects' rules of proportion, should have been rediscovered in the lay world of princes, scholars, and gentlemen, and should have led to that Renaissance of architecture in Italy that swept away Gothic once and for all. Alberti's *Ten books*, completed in 1452 and published in 1485, reinforced the authority of Vitruvius, and stressed, above all, the importance of proportion for correct and beautiful building. Brunelleschi had made exact measurements of classical buildings in Rome; now, with the invention of printing, it became possible to publish an illustrated Vitruvius. With the rules in one hand and illustrations in the other, it is no wonder that an educated man could, successfully, try his hand at architecture without the trouble of actually learning how to build.

Printing also furnished another nail in the Gothic coffin with the wide dissemination of the secrets of the lodges. Roriczer's (1486) *The right way for pinnacles (Der Fialen Gerechtigkeit)*, for example, is a textbook for apprentices (Roriczer succeeded his father as cathedral architect at Regensburg). The exercise given is simple, but it reveals permanently those rules for construction of irrational lengths and for sub-division of the basic module that were at the secret heart of an apprentice's training. But the rules were, in any case, no longer needed – the men of science had acquired the decimal notation from the Arabs. The irrational value of the square root of 2 could now be measured to any degree of practical accuracy on a decimally sub-divided rule.

The lodges did not take kindly to science. They seem to have resisted stubbornly the infiltration of new ideas, to the point where the 'Gothic' carpenter in Britain, well into the second half of the twentieth century when the country adopted the metric system, was still measuring work in eighths of an inch or, if greater accuracy were needed, in sixteenths, sub-dividing the basic module in the way that had been current for two millenia. The lodges were content to copy their rules from generation to generation, and surviving writings from Gothic workshops, already apparently dead and stilted by the time of Mignot, seem to become increasingly corrupt.

Rodrigo Gil de Hontañon's rules for the abutments of bridges and buttresses for vaults, of the first half of the sixteenth century, still involve geometrical constructions, but the sketches are not self-consistent, and it is difficult to avoid the conclusion that Rodrigo was copying mechanically something he had learned by rote. There is still no hint in any of these writings of any understanding of the equilibrium of forces – the rules continue to be purely geometrical. Blondel's rule of 1683 for the widths of piers of arches consists only of geometry (fig. 8.6). The arc $ACBD$ (of any of the four arches) is divided into three equal parts by the points C and D. DB is extended so that $DB = BE$. Then E defines the outer edge of the abutment.

Blondel's rule accords with intuition, at least for single-span bridges; shallow arches will give greater thrusts, and need thicker abutments, and a round arch may thrust less than a pointed arch. It was not until a

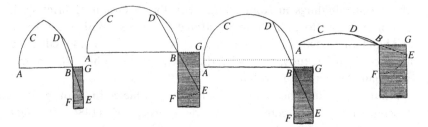

Fig. 8.6. Blondel's rule (1683) for the width of piers.

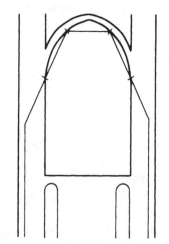

Fig. 8.7. Blondel's rule and the Sainte-Chapelle, Paris.

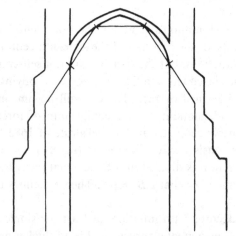

Fig. 8.8. Blondel's rule and King's College Chapel, Cambridge.

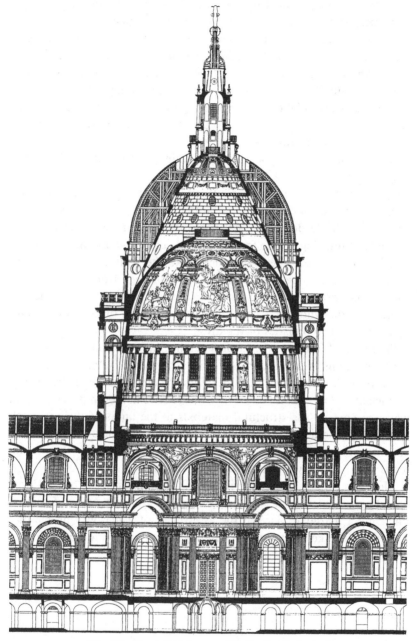

Fig. 8.9. The triple dome of St Paul's Cathedral.

century later that Perronet reduced drastically the sizes of the river piers of his bridges from those given by such medieval rules of proportion.

It is of interest to apply Blondel's rule to two buildings of roughly the same dimensions, the High Gothic Sainte-Chapelle of the thirteenth century, and King's College Chapel of the fifteenth. The cross-sections of the two buildings are shown to the same scale in figs. 8.7 and 8.8; the Sainte-Chapelle is slightly taller and King's College Chapel slightly wider (and has more bays). The later English work may be thought timid, if Blondel is a reliable guide.

By this time, of course, science was making its way outside the Gothic lodges. In 1675 Hooke knew of the hanging chain, and his co-Surveyor for London after the Great Fire, Wren, applied the idea spectacularly, and in a way inconceivable to a Gothic architect, in his design of St Paul's. The triple dome (fig. 8.9), consists of a timber outer structure (capable of resisting tensile forces), the true conical dome supporting the massive lantern, and the inner dome (like the Pantheon, with an eye) which is all that is seen from inside the building. Iron chains help to contain the thrust, but the main buttressing is provided by the inclined surfaces of the whole internal supporting structure, disguised externally by the vertical colonnades of the drum.

Before this, however, in 1638, as noted in Chapter 1, Galileo was laying the foundations of modern structural analysis. His first 'new science' was based on rational mechanics rather than the empirical geometrical rules of ancient and medieval times. Once this science had been developed, the way was open to use structural materials to the limits of their strength, and to abandon design rules based upon geometry. The modern designer calculates the strength and the stiffness of his structures; he estimates the loads they can carry and the deflexions they will develop.

These are not the design criteria of the masonry structure. Correspondingly, modern structural calculations and modern codes of practice find only a marginal place in the assessment of a masonry bridge or a cathedral. The key to the understanding of masonry is to be found in a correct understanding of geometry.

Bibliography

References in the text

Abraham, P. (1934). *Viollet-le-Duc et le rationalisme médiéval*. Paris.

Ackerman, J.S. (1949). *Ars sine scientia nihil est*, Gothic theory of architecture at the Cathedral of Milan. *The Art Bulletin.* **31**, 84–111.

Beckmann, Poul (1985). Strengthening techniques. *The Arup Journal*, **20**, No. 3.

Beltrami, L. (1902). Fall of the Campanile of St Mark's, Venice. *Journ. RIBA*, 3rd Series, **9**, 26 July 1902; *ibid.*, 27 September 1902.

Benouville, L. (1891–2). Étude sur la Cathédrale de Beauvais. *Encyclopédie d'architecture*, **4**, 4th series, 52–4, 60–2, 68–70.

Blasi, C. and Ceccotti, A. (1984). La cupola del Brunelleschi: Indagine sullo stato di fessurazione. *Ingegneri Architetti Costruttori*, June/September.

Blondel, Francois. (1675, 1683). *Cours d'architecture*, 2 vols. Paris.

Caroe, Alban D.R. (1949). *Old churches and modern craftsmanship*. Oxford University Press.

Choisy, A. (1883). *L'art de bâtir chez les Byzantins*. Paris.

Choisy, A. (1899). *Histoire de l'architecture*, 2 vols. Paris.

Clouston, R.W.M. (1970). *Movements in church structures* (a four-page note to the Council for the Care of Churches).

Conant, K.J. (1944). Observations on the vaulting problems of the period 1088-1211. *Gazette des Beaux-Arts*, 6th series, **26**, 127–34.

Desjardins, G. (1865). *Histoire de la Cathédrale de Beauvais*. Beauvais.

Fitchen, J. (1955). A comment on the function of the upper flying buttress in French Gothic architecture. *Gaz. Beaux-Arts*, **45**, 69.

Fitchen, J. (1961). *The construction of Gothic cathedrals*. Oxford.

Frankl, P. (1960). *The Gothic*. Princeton.

Frost, A.J. (ed.). (1990). *Towers and bells*. The Central Council of Church Bell Ringers.

Galileo, Galilei Linceo. (1638). *Discorsi e Dimostrazione Matematiche, intorno à due nuove scienze Attenenti alla Mecanica & i Movimenti Locali*. Leida. (*Dialogues concerning two new sciences*, translated by H. Crew and A. de Salvio, New York, 1952.)

Gil de Hontañon, Rodrigo (*c.* 1500–1577), Insertion in *Compendio de Arquitectura* ...(Simón Garcia, 1681) edited and published in *El arte in España*, **7**, Madrid, 1868.

Guadet, J. (n.d.). *Éléments et théorie de l'Architecture*, 4 vols. Paris.

Harvey J. (1958). Mediaeval design. *Transactions of the Ancient Monument Society*, New series, **6**, 55–72.

Harvey, W. (1925). *The preservation of St Paul's Cathedral and other famous buildings*. London (The Architectural Press).

Hooke, R. (1676, *sic*, actually 1675). *A description of helioscopes, and some other instruments*. London.

Jackson, T. G. (1906). *Reason in architecture*. London.

Jackson, T. G. (1915). *Gothic architecture in France, England, and Italy*, 2 vols. Cambridge.

Jackson, T.G. (1913, 2nd edition 1920). *Byzantine and Romanesque architecture*, 2 vols. Cambridge.

Jackson, T.G. (1921). *The renaissance of Roman architecture*, 3 vols, Cambridge.

Mackenzie. F. (1840). *Observations on the construction of the roof of King's College Chapel, Cambridge*. London (John Weale).

Moseley, H. (1843). *The mechanical principles of engineering and architecture*. London.

Navier, C.L.M.H. (1826; Saint-Venant, Paris, 1864, 3rd critical edition.) *Résumé des leçons données à l'École des Ponts et Chaussées...*, Paris.

Poleni, G. (1748). *Memorie istoriche della gran cupola del Tempio Vaticano*. Padova.

Rondelet, J. (1834). *Traité théorique et pratique de l'art de bâtir*, 7th edition (5 volumes + plates). Paris.

Roriczer, Matthaus. (1486). *Puechlein der Fialen Gerechtigkeit*. Regensburg.

Sabouret, V. (1928). Les voûtes d'arêtes nervurées. *Le génie civil*, 3 March.

Schonland, Sir Basil. (1964). *The flight of thunderbolts* (2nd edition). Oxford (Clarendon Press).

Steel Structures Research Committee, *First Report*, 1931, *Second Report*, 1934, *Final Report*, 1936, London (HMSO).

Ungewitter, G. (1901). *Lehrbuch der Gotischen Konstruktionen*. 4th edition (neu bearbeitet von K. Mohrmann), 2 vols, Leipzig.

Villard de Honnecourt. (*c.*1235). *Lodge book*. See (1) R. Willis (translated and edited by), *Vilars de Honecourt, Facsimile of the sketch-book of, with commentaries by J.B.A. Lassus and J. Quicherat*, T.H. and J. Parker, 1859. (2) H.R. Hahnloser, *Villard de Honnecourt*, Vienna, 1935, 2nd edition Graz, 1972.

Viollet-le-Duc, E.E. (1858–68). *Dictionnaire raisonné de l'architecture française du XIe au XVIe Siècle*, 10 vols. Paris.

Vitruvius. (*c.* 30 BC). *De Architectura*.

Willis R. (1842). On the construction of the vaults of the Middle Ages. *Trans. RIBA* **1** part II, 1.

Willis, R. (1861). *The architectural history of Chichester Cathedral*. Chichester.

Yvon Villarceau. (1854). L'établissement des arches de pont. *c.r. Acad. Sci. Paris, Mémoires présentés par divers savants*, **12**, 503 (also published separately 1853).

Publications by the Author

Coulomb's memoir on statics: An essay in the history of civil engineering. Cambridge University Press, 1972.
Equilibrium of shell structures. Oxford University Press, 1977.
The masonry arch. Chichester (Ellis Horwood), 1982.

The stone skeleton, *Int. J. Solids Structures*, **2**, 249, 1966
On shell solutions for masonry domes, *Int. J. Solids Structures*, **3**, 227–41, 1967.
Spires and fan vaults, *Int. J. Solids Structures*, **3**, 243–57, 1967.
Westminster Hall Roof, *Proc. Instn civ. Engrs*, **37**, 137, 1967.
On the rubber vaults of the Middle Ages, and other matters, *Gaz. Beaux- Arts*, **71** 177, 1968.
The safety of masonry arches, *Int. J. Mech. Sci.*, **11** 363, 1969.
Beauvais Cathedral. *Trans Newcomen Soc.*, **40**, (1967–68), 15, 1971.
'Gothic' construction in Ancient Greece, *J. Soc. Architect. Hist.*, **31**, No. 1, 3, 1972.
Two masonry bridges: I. Clare College Bridge, *Proc. Instn civ. Engrs*, **52**, 305, 1972 (with C.J. Padfield).
Two masonry bridges: II. Telford's bridge at Over, *Proc. Instn civ. Engrs*, **52**, 319, 1972 (with B.D. Threlfall).
The strengthening of the West Tower of Ely Cathedral, *Proc. Instn civ. Engrs*, **60**, 123, 1976.
Couplet's engineering memoirs, 1726–33, in *History of technology 1976* (ed. A. Rupert Hall and Norman Smith), (Mansell), 1976, pp. 21–44.
Inertia forces due to bell-ringing, *Int. J. Mech. Sci.*, **18**, 161, 1976 (with B.D. Threlfall).
An apsidal timber roof at Westminster, *Gesta*, **15**, 53, 1976.
The Gothic structure, *Interdisc. Sci. Revs*, **2**, 151, 1977.
The restoration of masonry: structural principles, *Architect. Sci. Rev.*, **20**, no. 2, 35, 1977.
The rehabilitation of Teston bridge, *Proc. Instn civ. Engrs*, **68**, 489, 1980 (with N.B. Hobbs and B.S. Jermy).
The estimation of the strength of masonry arches, *Proc. Instn civ. Engrs*, **69**, 921, 1980.
Shibam and Wadi Hadramawt, Report No. 3, Serial No.FMR/CLT/CH/82/139. Technical Report RP/1981–1983/4/7.6/04, UNESCO, Paris, 1982 (with R. Lewcock).
La restauration des ouvrages de maçonnerie: principes structurels, in *Restauration des ouvrages et des structures*, Paris (Presses de l'école nationale des ponts et chaussées), 1983.
Chronic defects in masonry vaults: Sabouret's cracks, *Monumentum*, **26**, 131, 1983.
The high endurance of the masonry structure, in *Durability and design life of buildings*, The Institution of Civil Engineers, 1984, pp. 59–65.
Calculation of abutment sizes for masonry bridges, in *Colloquium on History of Structures* (International Association for Bridge and Structural Engineering), 1982 (Institution of Structural Engineers, 1984).

The maintenance of masonry; papering over the cracks, in *Proceedings of the Symposium on Building Appraisal, Maintenance and Preservation*, Bath, July 1985 (The Institution of Structural Engineers).

The crossing piers of the French Pantheon, *Struct. Engr.*, **63A**, 230, 1985.

The timber octagon of Ely Cathedral, *Proc. Instn civ. Engrs*, **78**, 1421, 1985 (with E.C. Wade).

Statical aspects of masonry vaults and domes, in *Restoration of Byzantine and Post-Byzantine monuments*, Thessaloniki, 1986, pp. 229–35.

Masonry arches, vaults, and domes, in *Encyclopedia of Building Technology* (ed. Henry J. Cowan), (Prentice-Hall), 1988.

The structural analysis of Gothic architecture, *Proceedings of the Royal Institution*, **59**, 215–26, 1987.

Poleni's problem, *Proc. Instn civ. Engrs*, **84**, 737, 1988.

The care of masonry buildings: the engineer's contribution, in *Structural repair and maintenance of historical buildings* (ed. C.A. Brebbia), Southampton (Computational Mechanics Publications), 1989, p. 3.

Hemingbrough Spire, in *Structural repair and maintenance of historical buildings II, 1: General studies, materials and analysis* (ed. C.A. Brebbia, J. Dominguez, F. Escrig), Southampton (Computational Mechanics Publications), 1991, p. 13.

How to design a cathedral: some fragments of the history of structural engineering, *Proc. Instn Civ. Engrs Civ. Engng*, **92**, 24–9, 1992.

Leaning towers, *Meccanica*, **27**, 153–59, 1992.

The collapse of stone vaulting, in *Structural repair and maintenance of historical buildings III* (ed. C.A. Brebbia and R.J.B. Frewer), Southampton (Computational Mechanics Publications), 1993, pp. 327–38.

The roof of the monks' dormitory, Durham, in *Engineering a cathedral* (ed. M. Jackson), London (Thomas Telford), 1993, pp. 169–79.

Index

Abraham, Pol 66f., 88f.
Ackerman, J.S. 2
ad quadratum, ad triangulum 148
Amiens cathedral 51, 103ff.
Anthemios 2, 5
arch, masonry
  collapse 4, 18f.
  flat arches infinitely strong 20
  geometrical factor of safety 20ff.
  hinges in 4, 15ff.
  line of thrust 7, 17f.

Basilica of Maxentius 51
Bologna, S. Petronio 26
boundary conditions 8, 9
Brunelleschi 26, 27, 42, 46f., 150

Cambridge, King's College Chapel 78, 79, 81f., 91f., 152, 154
chain, hanging 7, 18, 28, 36, 81f., 154
Choisy, A. 3, 43ff.
cracking of masonry 15ff., 23, 35f.
crossing piers 13, 24, 116, 118
crushing strength 12, 13, 30

dome 27ff.
  cracking 35f., 46f.
  half dome as buttress 43
  hoop stresses can be tensile 33
  low stresses 30
  minimum thickness 39ff.
  Poleni's analysis 35ff.
  as a shell 27f.
drift of masonry 84f., 87, 110f., 119

earthquakes 1, 4, 71
Ely, Cathedral 25, 110, 119ff.
Ezekiel 141f.

fire, effects of 71f.
Florence, Brunelleschi's dome 27, 42, 46f.

flying buttress 91ff.
  failure at Amiens 103f.
  as flat arch 20
  passive and active states 98ff.
  position of 58
  to resist wind 24, 91f.
  Ungewitter's analysis 95ff.
foundations, movement of 1, 15, 23, 46, 49, 68, 110, 118f.
friction forces 13, 90

Galileo 5, 26, 154
Gaudí, A. 95
Gaussian curvature 76
geometrical factor of safety 20ff.
great measure 141f., 148f.
Guadet, N. 91

Hagia Sofia, Constantinople 1, 2, 5, 27, 41, 43ff.
Harvey, J. 2
hinging cracks
  in arches 16ff.
  in vaults 49
Hooke, Robert 7, 18, 28, 36, 154

irrational numbers 142
Isidorus 2, 5

lightning, effects of 110f.
Lincoln, Chapter House 103

mechanism of collapse 4, 18f.
Mignot 149f.
Milan cathedral 2, 88, 148ff.

Navier, C.L.M.H. 6, 8

Pantheon, Rome 4, 28, 29, 50
Paris, Notre-Dame 75, 91, 101f.
Paris, Sainte-Chapelle 91, 152, 154
Paris, Sainte-Geneviève 86f.
pendentives 27

159

Printed in the United States
By Bookmasters